機械を説明する英語

boring【中ぐり】
counter sinking【皿座ぐり】
drilling, drill【穴あけ】
spot【(浅い)穴をあける】
control【制御する, 操る】
handle【扱う, 取扱う, 操作する, 処理する】
manage【扱う, 操作する, 管理する】
manipulate【扱う, 操作する】
operate【操る, 操作する, 動かす】
take【扱う】

野澤義延 著

ease【徐々に動く】
go【通る, 移動する】
move【動く, 動かす, 移動する, 転送する】
travel【移動する, 走行する(行程, 移動量)】
traverse【移動する, 横行する】
complete【してしまう, 完了する, 仕上げる】
end【(完結)する】
finish【してしまう, 仕上げる】
feed【送る, 送り量, 供給する】
fashion【(特定の)形にする】
form【(形)づくる, 形成, 成形】
mould(mold)【(形を)つくる, 成形, 成型, 鋳込む】
shape【(形を)つくる】
attention【気をつける(配慮)】
care【気をつける】
caution【警告する, 注意する】
note【注意する】
precaution【用心する, 予防策】
chip load【切り込む】
cut into【切り込む】
depth of cut【切り込み】
engage【切り込む, 噛合う】
broaching【ブローチ削り】

森北出版株式会社

まえがき

英語に強くなくても，専門のことなら，レポートは図，式，数字，記号で大体理解できるし，新しい論文で日本語化されていない用語にぶつかっても特別抵抗を感じることはあまりないと言っていい．ただし，それは「英」→「和」の場合である．

今は，日本の製品を英語で紹介することが当たり前という時代で，輸出指向の企業ではずっと前から英語のパンフレットや，取扱い説明書は常識である．「和」→「英」となると，図と記号だけというわけにはゆかない．機械の動きや特徴を説明するために文章も必要になる．また，技術者が英文の取扱い説明書を作るように求められるケースも少なくない．単に専門用語の置換えではすまされず，製作（加工，組立），試験・検査，使い方（取扱い，操作，保全），機械のからくり（機能，動き）などを説明する英文が必要になる．

工業英語の用語集や手引き書は数多いが，この点を重点に扱った手頃なものは案外見当らない．本書の意図はここにある．したがって，これは「……する」とか「……になる」とか作業，動作，現象などを主調にした「和」語に対する英語の類語，類例集である．

本文では，見出しの英語に対して「英」で意味を引用した〔The Oxford Paperback Dictionary (1879)〕．これは類語の意味の近似性だけでなく，学術用語と基本語との関連を理解する上で，日本語よりも呑み込みやすいと考えたからである．学術用語と基本語のへだたり度（馴染みの度合い）は，機械工学の分野で英は26.5%，日本は55.5%という〔角川：図説日本語〕．

このことは，専門用語が「和」より「英」のほうが日常の言葉にずっと近いということであるし，技術関係の説明もそう特殊な言い表わし方をしているわけでもないことを示している．とは言え，機械分野の英文はそれなりに慣用，多用されるものがある．「英」の例文は機械関係の専門書，論文，雑誌，カタログ，解説書，広告からの引用である．

本書が機械英語に馴染む上で，またカタログなどの作成時において一助となれば幸いである．

1986年11月　　野澤　義延

【追補】

本書は，1986年に工業調査会から発行したものを，弊社から継続して発行しました．

2011年6月　　森北出版

目　次

まえがき …………………………………………… 1

あける，くる，もむ …………………………………………… 15

boring : 中ぐり　15
counter sinking : 皿座ぐり　16
drilling, drill : 穴明け　17
spot : (浅い) 穴をあける　18
spot facing : 座ぐり　19

扱う，操る，こなす，処理する …………………………………… 19

control : 制御する，操る　19
handle : 扱う，取扱う，操作する，処理する　20
manage : 扱う，操作する，管理する　25
manipulate : 扱う，操作する　25
operate : 操る，操作する，動かす　27
take : 扱う　28

移動する，動く，移す …………………………………………… 28

ease : 徐々に動く　28
go : 通る，移動する　29
move : 動く，動かす，移動する，転送する　29
travel : 移動する，走行する (行程，移動量)　33
traverse : 移動する，横行する　35

終わる，してしまう (終了，完了，完結) ………………………… 36

complete : してしまう，完了する，仕上げる　36
end : (完結) する　37
finish : してしまう，仕上げる　37

送る …………………………………………………………………… 37

feed：送る，送り量，供給する 37
traverse：送る 41

形づくる（成形，整形，細工） ………………………………………… 42

fashion：（特定の）形にする 42
form：（形）づくる，形成，成形 42
mould(mold)：（形を）つくる，成形，成型，鋳込む 44
shape：（形を）つくる 45

気をつける（注意，留意，用心，配慮，考慮） ………………… 45

attention：気をつける（配慮） 45
care：気をつける 47
caution：警告する，注意する 48
note：注意する 49
precaution：用心する，予防策 49

切り込む ……………………………………………………………… 49

chip load：切り込む 50
cut into：切り込む 50
depth of cut：切り込み 50
engage：切り込む，噛合う 50

切る，削る，切削する ………………………………………………… 51

broaching：ブローチ削り 51
contour：輪郭削り 52
cut：切削する，機械加工する，削る，切る，カットする，切断する 53
facing：面削り 57
groove, grooving：溝削り 58
knurling；knurl：ローレット切り 58
machine：切削する，機械加工する（加工する），削る 59
milling；mill：フライス削り 64
necking：逃げ溝削り 66
profile：ならい削り，縦断面図 66
recessing, recess：逃げ溝切り 66
remove：取り去る，切削する，研削する，加工する 66
rough：荒削り 68

screw：ネジを切る　69
shaping：形削り　70
surfacing; surface：サーフェーシング，平面を削る　70
thread：ネジを切る　70
turning; turn：旋削する，外丸削り　71
under cut：逃げ切り　73

組立てる ……… 74

assemble：組立てる　74
build：（組立てて）つくる　76
combine：組合せる，一緒にする　76
connect：つなぐ　77
construct：（組立てて）つくる　77

けがく ……… 78

layout：けがく，段取りする　78
mark：けがく　78
scribe：けがく，けがき針　79

研削する，とぐ ……… 80

cut：削る　80
grind：研削する（とぐ）　80
remove：研削する　82
sharpen：とがらす，とぐ　82

合成する，一緒にする ……… 82

combine：合わせてつくる（結合，混合）　82
synthesize, synthesis：組合せる，合成する　84

こする，する（摩擦） ……… 84

chafe：こする，すりむく，擦過，摩擦　84
friction：摩擦　84
rub：こする，こすりつける，摩擦する　87

固定する，止める（取付ける） ……… 88

bolt：ボルトで固定する　88
clamp：締付ける，しっかり取付ける，クランプで固定する　88

fasten：固定する，取付ける，縛る 90
fix：固定する，しっかり取付ける 91
key：キーで固定する 92
lock：ロックする，止める，固定する 92
screw：ネジで固定する 94
seat：（台座などに）止め付ける，固定する，据える 95
secure：固定する，取付ける 95

さがす（捜査，探査），調べる …… 97

research：調べる（調査） 97
search：調べる，さがす（探索） 97
seek：さがす（探求） 97

支える，支持する，担う …… 98

bear：担う，支える 98
carry：担う，支える 98
prop：（棒で）支える 99
support：支える，支持する 99
sustain：支える 101

差し込む，取付ける …… 101

insert：差し込む，間に取付ける 101
load：取付ける，装塡する，積む 102
plug in：差し込む，取付ける 104

仕上げる …… 105

burnishing：バニシ仕上げ 105
dress：仕上げる 105
file：ヤスリ仕上げ 105
finish：仕上げる 105
honing：ホーニング仕上げ 108
lapping：ラップ仕上げ，ラッピング 109
reaming：リーマ仕上げ 109
scraping：キサゲ仕上げ 110
shave：シェービング仕上げ 111
skive：スカイビング仕上げ 112

してみる，やってみる，試みる …… 112

attempt：やってみる，つと（努）める 112
try：してみる，試みる 112

処理する …… 113

attend to：処理する，留意する 113
cope with：うまく処理する，こなす 114
deal with：処理する，処置する，（問題として）取上げる 115
overcome：克服する，処理方法がわかる 115
process：処理する 116
treat：処理する，扱う 117

調べる（観察，看視，調査，走査，検査） …… 119

assay：調べる（試金） 119
delve：調べる，掘り下げる 119
examine：調べる，よく見る（検査，検出，点検，検討） 119
explore：調べる 120
inspect：調べる，詳しく見る（検査） 120
monitor：モニターする，監視する 124
observe：調べる（よく見る，観察） 125
probe：調べる（探査，探測，探針） 125
read：調べる（点検） 126
scrutinize：詳しく調べる（精査，吟味） 126
see：調べる 126
study：調べる（研究） 126
survey：調べる（実地踏査，測量，検査） 127
test：調べる（検定，検査，試験） 127

する，やり遂げる（実行，完遂） …… 132

accomplish：する，やり遂げる，完遂する 132
apply：～をする 134
effect：（～を）する，してしまう 134
implement：する，実行する，具体化する 134

（～を）する（実行，実施） …… 135

carry out：（～を）する，実行する 135
conduct：（工夫して）する，処理する 136

do：（～を）する　137
execute：する，実行する　138
make：する　139
perform：する　143
practice：する，実行する　146
take：する，処理する　146
undertake：する　147

確かめる（確認）……147

analyse：調べる（解明，解析，分析）　147
check：調べる，確かめる（点検，検査）　148
validate：確かめる（確認）　151
verify：調べる，確かめる　151

たてる……153

tapping：タップ立て，ネジ立て，切り付け　153

調整する，（～を）合わせる，調節する，整える……154

adjust：調整する，加減して合わせる　154
control：制御する，管理する，調節する　160
dial：ダイヤルで調節する　164
focus：焦点を合わせる　164
regulate：調整する，調節する　164
set：合わせる，セットする　165
tune：調子を整える　166

ついや（費）す……166

consume：使い尽す，消費する　166
expend：ついやす　167
spend：使う，ついやす　167
take：使い尽す（消費する）　167
use up：使い尽す　167
waste：むだに使う（消耗する）　168

つかう，使う，用いる……168

adopt：使う，採用する　168
apply：使う，適用する，応用する　169

employ：使う　171
make the most of：活用する　174
service：使う，役立てる，役に立つ，扱う　174
share：共用する　175
take：使う，利用する　175
take advantage of：使う，利用する，活用する　175
use：使う，役立てる　176
utilize：使う，活用する，利用する　184
with：使って，～で　187

つかむ，取る，にぎる，つまみ上げる ……………… 188

chuck：チャック，つかむ　188
clutch：しっかりつかむ　188
grab：つかむ　189
grasp：つかむ，把握する　189
grip：つかむ，にぎる，把持する　190
hook：（鈎で）ひっかけてつかむ　194
pick up：つまみ上げる，取上げる　195
snatch：ぐっとつかむ　196
take：取る　196
trap：つかまえる，補捉する　196

突切る ……………… 197

cutting off：突切る　197
part：突切る　198
sever：突切る　199

つく（創）る（創造，考案） ……………… 199

build：つくる，組立てる　199
construct：（組合せて）つくる　200
create：つくる（創る）　201
erect：つくる（建造する）　202
generate：つくり出す　202
invent：考え出す，発明，案出　203
make：つくる　203
produce：つくる　203

つくる（構成） ………………………………… 204

formulate：つくる 204
make up：つくる，組立てる 204

つくる（作成，製作，製造，生産） ………………… 205

fabricate：つくる 205
make：つくる 206
manufacture：つくる（製造，生産） 209
produce：つくる（製作，生産） 209
turn out：（加工して）つくる 213

吊す，懸ける，掛ける ………………………………… 214

hang：吊す 214
suspend：吊す 214

（～で）できている，（～で）構成されている ……… 215

built-in：組込む，内蔵している 215
compose：形成する，合成する 215
comprise：成り立っている，構成されている 216
consist of：構成されている，（～で）できている 218
constitute：構成する，（構成成分は）～である 219
contain：もっている，中にある 220
equip：装備している，付いている 221
feature：特徴ある構成要素，特徴である，～がある 223
form：形づくる 227
furnish：備え付ける 227
have：もっている，～がある 228
include：（～の中に）ある，～もある，付いている 229
incorporate：～もある，取り入れる，組み入れる 232
involve：含む，～もある，伴う 233
make up：つくる，形成する，構成する 234
possess：もっている 236
provide：設けてある，付いている，用意されている 236
supply：供給する，設ける 238
with：付いた，持った 238

できる，できない，してもよい（可能，能力，余裕） ……… 239

ability：できる，能力，資質 239
able：できる 239
allow：できる，してもよい 239
can：できる 243
capable：できる，能力がある 244
capacity：できる（容量，能力） 245
enable：できる，手段を与える 246
fail：できない，うまくいかない 247
feasible：（することが）できる 248
give：できる 249
may（might）：できる，してもよい 249
permit：できるようにする 249
possible：できる 252
provide：できる，～が得られる 254

ととのえる（調製，用意，準備） ……… 255

improvise：（即席で）つくる 255
prepare：つくっておく，調製する 255

取る（面を） ……… 256

chamfering；chamfer：面取り，食付き部（タップ，チェーサ） 256
round：丸面取り，丸める 257

取付ける，置く，降ろす，乗せる ……… 258

attach：取付ける 258
deposit：置く，降ろす 259
fit：取付ける，はめる 260
fit up：設ける，取付ける 262
install：取付ける，据付ける 263
lay：置く 264
locate：位置決めする，～にある，取付ける 264
mount：取付ける 265
place：置く，取付ける 269
position：置く，位置決めする 272
put：置く，取付ける 272
rest：置く，静置する 273

reset：取付け直す，リセットする　274
ride：乗る　274
set：取付ける，据付ける，設定する　275
set up：取付ける，アレンジする，段取りする　277
situate：置く，取付ける　278

取外す …… 278

detach：取外す　278
disconnect：取外す，連結を解く　279
dismount：取外す　280
remove：取外す　280
take away：取外す　282
take off：取外す　282
withdraw：取外す，引き戻す　282

はかる，測る，計る，量る，秤る …… 283

determine：定量する，正確に求める，はっきりさせる　283
gauge(gage)：測る　285
measure：測る，測定する，計測する，寸法は～である　286
meter：測る，計量する，量は～である　294
pick up：測る　295
read：読む，読取る，読み出す，示す　295
scale：測る　296
size：寸法決めする　296
survey：実測する，測量する　297
take：～を測る　297
weigh：（重量を）計る，秤る，重さは～である　297

はかる，図る，もくろむ（計画，設計，立案） …… 299

arrange：計画をつくる　299
contrive：上手に計画する　299
design：計画する，設計する，デザイン　299
engineer：設計する，処理する　303
plan：計画する　303
program(programme)：プログラムを組む，プログラム，実行可能プログラム，ルーチン　304
schedule：計画する，日程計画，（指定）工程　305

挟む，くわえる ………… 306

nip：挟む，くわえる　306
pinch：挟む，つねる　306

ひく ………… 307

sawing：鋸引き　307

分解する ………… 307

break up：分解する，解体する　307
disassemble：分解する　307
dismantle：分解する　308
take apart：分解する　308

磨く ………… 309

polish：磨く，研磨，研削　309
sanding：磨く，研磨　310

見付け出す（検知，検出） ………… 310

ascertain：調べる（検知）　310
detect：見い出す（発見，検出，検知）　311
determine：調べる　311
trace：追跡調査する　312
trouble shoot：調べて直す（故障探求）　313

持つ，保つ，保持する，維持する ………… 313

hold：持つ，つかむ，支える，保つ　313

〈Index〉 ………… 317

あける，くる，もむ

boring: to make (a hole or well etc.) with a revolving tool or by digging out soil　　中ぐり

You will be able to use an offset boring head to **bore** a hole pattern to the specified dimensions and within the tolerances given.

オフセット・ボーリングヘッドは，穴形状を与えられた公差内で仕様寸法に**中ぐり**するのに利用できる．

Twice as fast : Designed to take advantage of today's higher horsepower machine tools and machining centers, Microbore Twin-Bore Tooling can rough **bore** at feed rates twice as fast as conventional tooling. The reason is simple. The use of two cutting edges instead of one.

2倍の速さ：今日の高馬力工作機械やマシニングセンタを活かすように設計されたMicrobore ツインボア・ツーリングは，普通のツーリングの2倍の送り量で荒**中ぐり**できる．理由は簡単．切れ刃が1つでなく，2つ使用だから．

We use this tool to cut oil grooves in bronze bushing in an engine lathe. The basic member is a flanged disc arranged for mounting to the lathe face plate. This disc **is bored out** to take the bushing, and its outside diameter is threaded to accommodate a ring that locks the bushing in place.

この工具は，普通旋盤で青銅のブシュに油溝を切るために使う．基本部品は，旋盤の面板に取付けられるようにアレンジしたフランジ付円盤である．この円盤は，ブシュを取り入れるように**くり抜かれ**，その外径は，ブシュを所定の位置に固定するリングが受入れられるように，ネジを切ってある．

Cutting tool (a) **counterbores** engine blocks, and sensing element (b) of electronic control maintains the required depth.

バイト(a)でエンジンブロックを**深座ぐり**し，電子制御の検出素子(b)で所要の深さを維持する．

Explain the differences between through **boring**, **counter boring** and **boring** blind holes.

通し**中ぐり**，**深座ぐり**，めくら穴**中ぐり**の違いを説明せよ．

Boring is the process of enlarging and truing an existing or drilled hole. A drilled hole for boring can be from 1/32 to 1/16 in. undersize, depending on the situation. Speeds and feeds for **boring** are determined in the same way as they are for external turning. **Boring** to size predictably **is** also **done** in the same way as in external turning except that the cross feed screw is turned counterclockwise to move the tool into the work.

中ぐりは，既存あるいは穴明けした穴を拡大して正しいものにする工程である．**中ぐり**用に穴明けする穴は，状況にもよるが，1/32～1/16インチ以下のサイズでよい．**中ぐり**のスピードや送りは，外径旋削のそれと同じやり方で決める．また，想定寸法に**中ぐりする**ときも，バイトを工作物の中へ動かすために横送りネジを反時計方向に回す以外は，外径旋削と同じやり方で**行なう**．

The most common machining process is boring and is shown in the figure ; **boring is achieved** by rotating the tool, which is mounted on a boring bar connected to the spindle (motion C), and feeding the spindle, boring bar, and tool along the axis of rotation (motion Z).

中ぐりは最も一般的な機械加工法で，図の通り：**中ぐりは**，主軸に連結し片持中ぐり棒に取付けてあるバイトを回転し（運動C），回転軸に沿って主軸，片持ち中ぐり棒およびバイトを送ること（運動Z）で**行なわれる**．

〈用　語　例〉

boring bar tool　中ぐりバイト
boring head　中ぐり刃物頭
boring table　中ぐり台
boring bar　中ぐり棒
boring tool　中ぐりバイト
counter boring　深座ぐり
stub boring bar　片持中ぐり棒
wood boring lathe　穴明け旋盤

counter sinking ; counter sink : to dig (a well) or bore (a shaft)　**皿座ぐり**

Counter boring is a process related to drilling and is employed to form a

深座ぐりは，穴明けに関連した加工工程で，既存の穴の端部

cylindrical hole of large diameter at the end of an existing hole——e.g., to receive the head of a screw or bolt. If the enlarged hole is formed with tapered sides, the process is called **countersinking**.

から，たとえばネジあるいはボルトの頭を入れるために，より径の大きい円筒形状に加工する場合に利用される．もし，拡大した穴がテーパの側面で形成されると，その加工法を**皿座ぐり**という．

〈用　語　例〉

countersink　皿穴
countersinking　皿もみ
counter sunk square head　角皿(頭)(ネジ)
counter sunk square neck　角根丸(頭)(ネジ)
oval counter sunk head　丸皿(頭)(ネジ)

drilling : to make a hole with a drill
drill : a pointed tool or machine used for boring holes or sinking wells)　**穴明け**

With proper attachment and proper adjustments, a lathe can **drill**, ream, tap and thread.

適切な治具と適切な調節により，旋盤で**穴明け**，リーマ仕上げ，タップ立て，ネジ切りができる．

Deka Drill simultaneously **drills** 14 anchor nut forgings in 15 seconds. A second machine taps the holes in 10 seconds.

Dekaボール盤は，15秒で14のアンカーナット鍛造品を同時に**穴明けする**．第2の機械は，10秒でその穴にタップを立てる．

The first **drilling** operations on the 11 station drilling machine involve two 6.68 mm diameter holes. These **are drilled** to half the required depth at one station and to full depth at the next station. A 6.45mm diameter tumbler hole **is** then **drilled** to full depth. These three holes **are** subsequently **redrilled** for burr removal.

11ステーション・ボール盤の**穴明け**作業には，最初直径6.68 mmの穴が2つある．これらは，最初のステーションで所要深さの1/2まで，つぎのステーションで全深さに**穴明けされる**．そして，直径6.45mmのタンブラ用穴が全深さまで**穴明けされる**．続いて，3つの穴は，バリ取りのために**再度穴明けされる**．

Use a 1/4in. diameter twist drill and **drill** this hole $1\frac{1}{2}$in. deep.

1/4インチ径のツイスト・ドリルを使って，深さ$1\frac{1}{2}$インチの穴を**あける**.

Center **drill** this hole.

この穴のセンタを**もむ**.

One of the conclusions drawn from these tests is that peck **drilled** holes have better surfaces than either STEM or EDM **drilled** holes.

これらテストから得た結論の1つは，ペック（ツツキ）**穴明けした**穴の表面はSTEMやEDM **穴明けの**穴より良い，ということである.

You can easily **laser drill** all metals as well as ceramics, cermets, saphire and diamond.

あらゆる金属，セラミックス，サーメット，サファイヤ，ダイヤモンドも，簡単に**レーザで穴明け**できる.

Heavy duty **drilling** should **be done** on an upright or radial drill press.

強力な**穴明け**は，直立またはラジアルボール盤で**すること**.

How **is** laying out and **drilling** center holes in a drill press **accomplished**?

ボール盤でのセンタ穴のけがきと**穴明け**は，どのように**する**か？

Now, with the Valenite Centre-Dex single or double-flute end mills, plunge **drilling** to a flat bottom **can be accomplished** without primary or secondary operations.

Valenite Centre-Dexの1溝または2溝のエンドミルを用いて，1次，2次作業なしに，平らな底にプランジ**穴明けができる**.

〈用　語　例〉

drilling machine　ボール盤　　spot drilling　スポット穴明け

drill key hole　ドリル抜き穴

spot: roundish mark or stain, to mark with a spot or spots　（浅い）穴をあける

A 90° spade drill can take place of different-sized spot drills. It can **spot** holes of many diameters, the diameters always being twice the depth drilled.

90°の鋤ドリルは，いろいろな寸法のスポット・ドリルの代わりになる．それは各種直径の**穴をあけることが**できるが，その直径は常に穴明け深さの2倍である．

spot facing: to finish a flat face　　**座ぐり**

Now you can economically **perform** an almost endless variety of drilling, reaming, counter-boring, **spot-facing**, chamfering and tapping operations with N's improved F. machines.

N社の改良したF機は，ほとんど無限に各種穴明け，リーマ仕上げ，深座ぐり，**座ぐり**，面取り，タップ立て作業が経済的**にできる**．

扱う，操る，こなす，処理する

control: to manage, to regulate　　**制御する，操る**

Tools that **are controlled** by hand…….

手で**操作する**道具は…….

I found the car a bit difficult to **control** at high speeds, so I took it to the garage to have the steering checked.

高速で車がちょっと**操縦**しにくかったので，ステアリングをチェックするため車を整備工場にもっていった．

This **control** can handle up to 18 axes simultaneously and **control** either a single robot or a large, complex work cell with multiple arms or auxiliary sensors.

この**制御装置**は，最高18軸まで同時に扱かえ，単一のロボットだけでなく，多くの腕や補助センサをもつ大形で複雑なワーク・セルをも**制御する**ことができる．

〈用　語　例〉

control drum　操作ドラム

control input, manipulated variable　操作量

(operator) control panel, key board　操作盤

remote control valve　遠隔操作弁

handle: to move with hands, to manage, to deal with: to discuss or write about	扱う，取扱う，操作する，処理する

Handle with care.

取扱い注意.

Use care in **handling** parts to avoid possible injury to fingers.

指を怪我しないように部品の**取扱い**には，気をつけよ.

……, if the rings **are handled** so as to avoid excessive stretching when mounting.

取付けるときに，リングを伸ばしすぎないように**扱えば**，…….

Handle the movable frame with a string.

1本の紐で，可動フレーム(枠)を**操作する**.

Handling must be done with tweezers.

必ず，ピンセットで**取扱うこと**.

Mechanized **handling** beween each unit is achieved by robot or conveyor devices of advanced design.

斬新な設計のロボットまたはコンベヤ装置によって，各装置間の**ハンドリング**が機械化されている.

Careful **handling** of the fragile tools, usually with tweezers, is essential to prevent breakage.

折れやすいツールはていねいに**扱う**ことが（通常，ピンセットで），破損防止にきわめて大切である.

Rough **handling** of chip detector and the electrical connector is the major cause of failure and of false indications in the caution plate.

チップ検出器およびコネクタの粗雑な**取扱い**が，警報板の中の故障および誤表示の主な原因である.

The automatic pallet change system will **handle** pallets up to 400 mm and weighing as much as 347kg.

この自動パレット交換システムは，400mm まで，重量347kg までのパレットが**扱える．**

The grinding and polishing machine gives a clean, smooth finish to small parts made of metal or other rigid materials. **Handles** parts as small as 32 mm long, and as thin as 2.3mm.

この研削・研磨機は，金属など硬い材料でできた小物部品を，きれいで滑らかに仕上げる．長さ32mm，厚さ2.3mm の小物部品までも**扱える．**

Thin materials can **be handled** by using vacuum cups made of an elastic material. The vacuum seals can grip tightly without damaging the material, which is released quickly when the vacuum is broken.

薄い材料は，弾性のある材料で作られた真空カップを使うことによって**ハンドリング**できる．この真空シールは，材料を傷めずに，しっかり把持することができ，真空を切れば速やかに放せる．

……. This leaves the operator's hands free part **handling.**

これで，作業者は部品の**取り扱い**に手を使わないですむ．

Loading, unloading, and reversing of disc type and shaft-type parts is fully automated with T's new CNC flexible **handling** system.

円板形および軸形の部品の取付け，取外し，反転はＴ社の新しい CNC 汎用**ハンドリング**・システムを使って全自動化されている．

The B pick and place system can perform a variety of material **handling** tasks including machine loading and unloading, assembly, and transfer.

ピック＆プレースシステムＢは，機械への取付け・取外し，組立および移送など，いろいろな材料の**ハンドリング**作業をすることができる．

One of the greatest problems in realizing the objective of complete computer-

完全コンピュータ制御の製造設備の目標を，実際に実現する

controlled manufacturing facilities is the development of general-purpose, programmable **handling** devices.

うえで最も大きな問題の１つは，汎用プログラマブル・**ハンドリング**装置の開発である．

This type of chip will fall into the chip pan and **is** more easily **handled**.

この種の切屑は，切屑受皿に落ちるから，より楽に**扱える．**

The cutters can **handle** the exotics, heatresistant alloys, and the ultrahigh-strength steels.

このカッタは，新種の耐熱合金および超高力鋼を**加工する**ことができる．

The company uses three drillheads to **handle** 75,000 pieces per year.

この会社は，年75,000個を**処理する**のに３台のドリルヘッドを使っている．

Several workpieces **are handled** in the plant simultaneously.

この工場では，数種の工作物を同時に**さばいている．**

The system cuts to within 0.004″ (on dia.) so grinders need only **handle** the finishing operations.

このシステムは，0.004インチ以内（直径で）で切削できるので，研削盤を必要とするのは仕上作業を**する**ときだけである．

The unit can **handle** the extraction of bushing, ……, from holes 1/4 to 25/32 in. in diameter.

この装置は，直径1/4から25/32インチまでの穴から，ブッシング，……の引き抜きを**する**ことができる．

General machining can **be handled by** a dilution of 20：1, while richer concentrates are recommended for slower speeds and heavier cuts.

一般的な機械加工には，20：1に薄めたもので**こなせる**．しかし，低速，重切削には，もっと濃度の高いほうがよい．

Manufacturing in the company's Space Systems Division **is handled** in a large facility, which includes more than one

この会社の宇宙システム部門の生産は，床面積が百万平方フィートを超えるような大きな施

million square feet of floor space.

設で**行なわれている．**

Programming is simple and can **be handled by** anyone familiar with machining operations after a brief training period.

プログラミングが簡単なため，機械加工作業に馴れた人なら誰でも，短期間のトレーニングで，**扱かう**ことができる．

The extremely valuable use of the computer is in the **handling** of complex mathematical computations.

コンピュータのきわめて価値ある使い道は，複雑な数学的計算の**処理**にある．

Model M-4 is built to **handle** high-speed machining work. 25 hp with spindle speeds up to 4,000 rpm. Tool changer capacity 40～100 tools.

M-4形は，25馬力で，4000 rpmまでの高速切削作業を**こなせる**ように作ったもので，ツール・チェンジャの容量は40～100ツールである．

Just add interchangeable tooling and you **handle** a wide variety of jobs with one D.

互換性のあるツーリングを付けるだけで，1台のDで多種の仕事を**こなせる**．

Bigger punching and shearing jobs can **be handled** by the Model 9, which has a 90-ton (801-kN) capacity.

これより大きな打抜きおよび剪断の仕事は，能力90t(801kN)を有する9形で**処理**できる．

The C can **handle** most pick and place tasks in the normal industrial environment, or can be used in the laboratory or classroom.

Cは，普通の産業環境条件での取ったり置いたりの作業のほとんどを**こなす**ことができ，研究所や教室で使える．

Standard carbide tooling is now available in a wide variety of types and styles to **handle** most machining applications.

標準品の超硬ツーリングは，広範多様な形式および様式のものが現在，利用可能で，切削加工のほとんどを**こなせる**．

A wide variety of everyday stamping

広範多種の普通のスタンピン

dies and similar applications **are handled by** low cost, two axis traveling wire electrical discharge machines.

グ・ダイおよびこれに類するものは，低コストな2軸移動形ワイヤ放電加工機で**処理**されている．

Liquids **handled** throughout all industries are of such volume and importance that can be described as liquid assets.

あらゆる産業の間で**使われている液は**，その量と，**重要さから**，**液体の資産**ということができる．

This lubricator can **handle** a wide variety of fluid and grease lubricants, including such T products as…….

この油差しは，……のようなT社製品など，広範多種の液およびグリース潤滑剤を**扱う**ことができる．

The turning operation is further complicated by an eccentric whipping motion of the part caused by its nonsymmetrical configutation. On the body for a 10″ ID valve, the diamond tool **handle** this condition and heat-hardened chunks of resin and overlapped fiberglass at a speed of 1,900 sfpm.

この旋削作業は，部品の非対称輪郭形状による偏芯むち打ち運動で，さらにややっこしくなる．内径10インチの弁本体について，ダイヤモンドツールは，樹脂上にガラス繊維を重ねた熱硬化の厚い塊りを，速度 1,900 sfpm で**処理する**ことができる．

The controller allows to **handle** occasional torque overload during brief periods of the test cycle.

この制御装置は，短期間の繰返し試験の間に，時として起こるトルクの過負荷に**対処する**ことができる．

In some situations a certain language may **handle** the problem more easily than another.

状況によっては，ある特定の言語がほかの言語よりも問題を楽に**処理**できることがある．

A manual jig grinder just couldn't **handle** the difficulty of the task.

手動のジグ研削盤では，この仕事のむずかしさを**こなし**きれなかった．

Experience has shown that this extrapolation can only **be handled** in a qualitative manner, rather than quantitatively.	周知のように，この外挿法は定量的ではなく，定性的にしか**扱え**ない．
J. Associates **handles** a number of tooling products, including L reamers, …….	J 社は，L 社のリーマ，……など数多くのツーリング製品を**扱っている．**
A number of measuring devices **are handled by** S. Among them is the D digital surface roughness gage.	S 社は，多数の測定装置を**取扱っている．**その中には，D 社の，ディジタル表面粗さ計もある．

〈用 語 例〉

handling　ハンドリング，荷役

handling time　ハンドリング・タイム，取扱い時間

manage: to operate (a tool or machinery) effectively ; to deal with, to be able to cope　**扱う，操作する，管理する**

A single operator can **manage** several machines or manufacturing cells.	たった 1 人の作業者で，数台の機械あるいは製造セルを**扱う**ことができる．
……, This appears to be the way towards the full implementation of CAE with its computer **managed** robot-operated factories.	……，これが，コンピュータ**管理**されたロボット稼動工場で，CAE（コンピュータ・エイデッド・エンジニアリング）完全実施への道であると思われる．

manipulate: to handle or manage or use (a thing) skillfully　**扱う，操作する**

The arm has the capacity to **manipulate** in six axes, and it has a motor for each	6 軸**操作**のこのアームは，各軸にはモータが 1 つ付いている．

axis. With the arm fully extended, it can handle loads of 0.45kg.

アームをいっぱいに伸ばした状態で，0.45kgの負荷を扱うことができる．

The part is inserted in the wash booth and the robot **manipulates** it through both washing and drying sequences.

部品が洗いブースに挿入されると，ロボットは連続した洗浄と乾燥作業でその部品を**扱う**．

The workpiece can **be manipulated** to present several sides to the spindles.

ワークは，いくつもの側面をスピンドルに向けるように**操作する**ことができる．

WING NUT,——A nut which possesses two "wings" or projections which enable it to **be manipulated** by the fingers instead of with a spanner.

蝶ナット．——2つの「ウィング」，すなわち突出部のあるナット．これによりナットはスパナを使わずに，指で**操作する**ことができる．

The question of repairs is seldom dealt with in typewriting manuals. It is possible for an experienced typist to make minor adjustments, but when a machine **is manipulated** with care there is usually little need for the services of the typewriter mechanic.

タイプライタの取扱説明書には修理の問題は，ほとんど採り上げられていない．機械を注意して**扱って**いれば，タイプライタ修理士による整備はまず必要なく，経験を積んだタイピストならちょっとした調整はできる．

It is because an electron carries an electric charge that it is particularly easy to **manipulate** by electrical methods.

電気的方法で**扱う**ことが特に容易であるという理由は，電子が電荷を帯びているからである．

The screw die cuts an external thread on a rod or bolt and is held in a device called a stock, which has handles for **manipulation**.

ネジダイスは，ナットやボルトに雄ネジを切るもので，**操作用**ハンドルの付いたストックという道具で保持する．

The multi-functional **manipulator** wrist

この多機能**マニピュレータ**の

has the ability to change its own hands or fingers and it can automatically take different tools it needs to perform certain tasks.

手首は，そのハンド部あるいはフィンガ部を取換えることができ，また特定の仕事をするために必要ないろいろなツールを自動で取ることができる．

〈用　語　例〉

manipulated variable　操作量，操作変量　manipulation　操作

operate : to control the function of, to be in action　**操る，操作する，動かす**

Simple to **operate**——No stop watches or thermometer required.

操作簡単——ストップ・ウォッチや温度計は不要．

Operate the handle carefully.

ハンドル**操作**はていねいに．

Tools can **be operated** by hand.

ツールは，手で**操作**できる．

The turret slide can **be operated** manually with the handle provided.

タレット送り台は，備え付けのハンドルを使って手で**操作**できる．

Sensitive drill presses are probably the most common machines used for drilling small holes. They **are** almost always hand **operated** with either a sliding quill to advance the rotating drill into the work, or an elevating table to force the work into the drill.

微動手送りボール盤は，小さい穴をあけるために使われる最も一般的な機械といえよう．これは，回転するドリルをワーク中に進めるために滑りクイルを，あるいはドリルにワークを押し入れるために昇降テーブルを使うが，通常はほとんど手で**操作する**．

Small, precision, **hand operated** turrets are used to produce very small parts.

小形で精密な**手動**タレットは，ごく小さい部品を作るために使われる．

Be sure that ailerons move in the correct direction when **operated** by the control wheel.

操縦輪で**操作**したとき，補助翼が正しい方向に動くことを確認すること．

Describe safe **operating** procedure for an NC machine tool.

NC工作機械の安全な**操作**手順について述べる．

Before discussing the **operation** of the lathe, it will be helpfull to describe …….

旋盤の**操作**について述べる前に，……に触れておいたほうがよいと思う．

〈用　語　例〉

logical operating　論理操作
off-line operation　オフライン操作
operand　オペランド，演算数
operating bar　操作バー
operating cost　操業コスト
operating device　操縦装置
operating fluid, hydraulic fluid　作動油
operating panel　操作盤
operating physical force　操作力
remote sequential operation　遠隔シーケンシャル操作
operating system　操作システム
operating time　動作時間，使用時間，演算時間
operation　操作，演算，運転，動作
operator (in symbols manipulation)　演算子
solenoid operated　ソレノイド操作

take: to perform, to deal with　**扱う**

The Model 300 will handle workpieces up to 300mm in diameter while the Model 630 **takes** workpieces up to 630 mm.

300形は直径300mmまでのワーク（工作物）が扱え，630形のほうは630mmまでのワークを**扱える**．

D's A-4 handles 6 or 12 channels, and B's D-2 **take** 24.

D社のA-4は，6または12チャンネルを処理でき，B社のD-2は24チャンネルが**扱える**．

移動する，動く，移す

ease: to move gently and gradually　**徐々に動く**

Start to fit the edge of tyre opposite the valve, pushing the edge into the well-base, and gradually **easing** the rest of the edge **into** place.

バルブの反対側のタイヤの縁から嵌め始め，この縁をウェル・ベースに押し込み，残りの縁を所定の位置に**徐々に入れる．**

go : to pass from one point to another
: (of a distance) to be traversed or accomplished

通る，移動する

The light **goes** from one substance into a less dense substance.

光は，ある物質からより密度の小さい物質に**移動する．**

At the end of an hour, if the trains continued to move uniformly, your train would have travelled 40km North, while the other would **have gone** 50km South.

列車が等速で動いているとすると，1時間後，君の列車は40 km 北に移動するが，もう一方の列車は50km 南に**移動している．**

move : to change or cause to change in position or place or posture ; to be or cause to be in motion

動く，動かす，移動する，転送する

The carriage will **move** the distance of the lead of the thread on the lead screw for each revolution of the lead screw.

往復台は，親ネジ1回転に対して，親ネジ上で親ネジリード分の距離を**動く．**

A distance **moved by** the tool…….

バイトが動いた距離は…….

Looseness allows the axle to **move** back-ward and forward for a short distance along the springs.

ガタがあるために，アクスルは，バネとともに短かい距離を前後に**動く．**

If the shock absorbers are correctly balanced each arm should **move** downward through the same distance in the same

ショックアブソーバのバランスが良ければ，アームはいずれも，同時に同じ距離を下の方向

time.

に動くはずである．

The lever **moves** through an angle of 100 degrees in a given number of seconds.

レバーは所定の秒数で，100°動く．

At the opposite end of the lathe bed from the headstock is the tailstock, which can **move** along the guideways and be clamped in any desired position.

主軸台より旋盤ベッドの反対の端にあるのが心押し台で，それは案内面に添って**動き**，どんな所望の姿勢にでも固定することができる．

The axle **move** vertically against the resistance of the spring.

アクスルは，バネの抵抗にさからって上下方向に**動く**．

The air **moves** from the high pressure region to the low pressure region.

空気は，高圧域から低圧域に**移る**．

The table **moves** quickly into position for grinding.

テーブルは，研削位置に素早く**動く**．

The table and saddle **move on** recirculating, pre-loaded ball cartridges **moving on** hardened and ground ways. **Movement** of table and saddle is by high-powered stepping motors which provide a rapid traverse rate of 200 ipm.

テーブルおよびサドルは，焼入れ研削した案内面の上を**動く**循環・与圧式ボールカートリッジに乗って**動く**．テーブルおよびサドルの**運動**は，強力ステッピングモータにより200ipmの早送りができる．

Each **moves** at a different speed.

それぞれ，異なった速度で**動く**．

A ball of radius, a, **moves** with a velocity, v, in a field of viscosity, η.

半径の a の球が，粘度 η の場を速度 v で**動く**．

The unit can handle workpieces from 3/4″to 8″ diameter and up to 11 lbs. per hand

このユニットは，グリッパを変えることによって，各ハンド

by changing the grippers. The system **moves** 800ipm on each axis and on two axes simultaneously.

は直径3/4～8インチ，11lbまでのワークを扱うことができる．システムは，800ipmで各軸および2軸同時に**動く**．

Punch presses with a basic cycle of 300 one-inch **moves** per minute are incorporating the AC servo.

動きが毎分1インチ300サイクル基本のパンチ・プレスには，ACサーボが付いている．

Precision ball screws **move** the table and saddle on ball cartridge.

精密ボールネジは，ボール・カートリッジ上のテーブルおよびサドルを**動かす**．

The MARC (Multi Axis Robotic Control) system, designed for automated stamping lines, is based on dual microprocessors to **move** and place a variety of metal sheet sizes.

このMARC（多軸ロボットの制御）システムは，自動打抜きライン用に設計されたもので，いろいろな寸法の金属板を**移動**して置くために，複式マイクロプロセッサを主制御装置としている．

Compress……by **moving** a position by *dx*.

*dx*だけ位置を**移動する**ことによって，……を圧縮する．

When a body **is moved** by the action of a force the force is said to do work.

物体がある力の作用によって**動かされた**とき，その力が仕事をするという．

The center sleeve in the tailstock can **be moved** in the longitudinal direction of the lathe by means of a handwheel and screw spindle and can thus be brought toward the workpiece.

心押し台のセンタースリーブは，ハンドホイールとネジスピンドルで，旋盤の縦方向に**動かす**ことができ，工作物の方向にもってゆくことができる．

The threading cutter is fed radially toward the workpiece, and either the

ネジ切りカッタは，工作物に向かって半径方向に送られ，カ

cutter or the workpiece **is moved** axially in synchronization with the slow motion of the workpiece.

ッタと工作物のいずれかが，工作物のゆっくりした回転運動と同期して，軸方向に**動く**.

When spraying the jet should **be** held about 4inch from the work and **moved** across the surface in straight lines.

吹付け塗装するときには，ノズルを工作物から約4インチ離して保持し，真直に表面を横切るように**動かすこと**.

The saddle can be traversed across the knee, and the work table can **be moved** to and fro in the guideways, by means of handwheels or by power drive.

ハンドホイールまたは動力駆動により，サドルはニーを横切って動かすことができ，ワークテーブルは案内面の中をあちこち**動かす**ことができる.

The drive screw which **moves** the table in and out is located beneath the center of the table….

テーブルを内外に**動かす**駆動ネジは，テーブルの中央下に位置し，…….

A simple test for looseness of the rear shackles is **to** grip the bumper and **move** the body up and down.

リア・シャックルのガタの簡単なテストは，バンパーをつかんで，ボディを上下に**動かす**ことである.

A head can **be moved** up or down the column of the machine.

ヘッドは，機械のコラムを上または下に**動く**ことができる.

The system uses an articulated arm to **move** pallets from a loading station to a horizontal work station situated on the table.

このシステムは，パレットを取付ステーションからテーブル上に置かれた水平の加工ステーションに**移す**のに，関節アームを使っている.

……no need to **move** the work from machine to machine.

……は，工作物を機械から機械へ**移す**必要がまったくない.

The overhead cranes **move** large castings and completed machines to various positions along the floor.

このオーバーヘッド・クレーンは，大形鋳造品および完成した機械を，床面に添っていろいろな位置に**移す**.

The steady rest **may be moved** to any location along the shaft.

固定振れ止めは，軸に沿ってどんな位置にも**動かせる**.

Move to the first hole location 1.015inch from the outside edge.

外縁から1.015インチの最初の穴位置まで**動かせ**.

Move the tool **off** the work and reset the cross feed micrometer to zero.

バイトを工作物から**離して**，前後送りマイクロメータをゼロにセットし直す.

It is better to have a lamp that can **be** easily **moved** about.

あちこち容易に**移動**できる照明具があったほうがいい.

〈用　語　例〉
moving-coil microphone　可動コイル・マイクロホン
moving coil pickup　可動コイル・ピックアップ
moving column type　コラム移動形

travel : to go from one place or point to another ; the range or rate or method of movement of a machine part

移動する，走行する（行程，移動量）

The tool **travels** a distance of 0.07 mm relative to the workpiece.

工具は，工作物と相対的に0.07mmの距離**移動する**.

The "pitch" of the thread is the distance, measured in the axial direction, between two corresponding points on adjacent turns of the thread ; it is the distance a nut would **travel** in one complete revolution.

ネジの「ピッチ」は，ネジの隣接するターン（巻き）の対応する2点間の，軸方向で測った距離である；これは，ナットが1回転で**移動する**距離である.

The Y axis spindle slide carrier **travels** along hardened and ground column ways at a rapid traverse rate of 400ipm.

Y軸のスピンドル・スライドキャリアは，早送り速度400ipmで，焼入れ・研削コラム面に沿って**移動する**.

Machine tables are accurately machined and the table **travels** parallel to its outside surfaces and also parallel to its T-slots.

機械のテーブルは正確に機械加工されていて，テーブルはその外側面と平行またはそのT溝と平行に**移動する**.

Unlike the grinding lathes, on this machine the workpiece **travelled** past the wheel instead of the wheel traversing the workpiece.

研削盤と違って，この機械では，砥石が工作物に沿って運動する代わりに，工作物が砥石に沿って**移動する**.

Chips **travel up** the drill flutes.

切屑はドリルの溝を**上がってゆく**.

Longitudinal **travel** is 30 inches per second. The arms **travel** up and down at a rate of 12.5″per second, and swing in and out in 40° per second.

左右の**移動**は毎秒30インチ．アームは毎秒12.5インチの速さで上下に**動き**，毎秒40°で内外に振れる.

One is moving upwards while the other **is travelling** downwards.

一方は上の方向に動いており，他方は下方向に**動いている**.

Vertical and horizontal **travel** of the head is electrically powered.

ヘッドの上下および水平方向の**移動**は，電動である.

Maximum horizontal **travel** is 2.25″ with adjustable stops. The vertical Z motion, also powered by a double acting pneumatic ram, has a maximum **travel** of 2.5″and is fully adjustable.

水平方向の最大**移動量**は，2.25インチ（可変ストッパ付）．また，復動空気圧ラムで動かす上下方向Z運動の最大**移動量**は2.5インチで，完全可調整である.

Vertical **travel** range is 18″.

上下方向の**移動**範囲は18インチである．

〈用　語　例〉

head travelling mechanism　主軸頭移動機構
over travel, over run　〔〔ホーンの〕オーバートラベル
travelling cable crane　走行ケーブルクレーン
travelling cut-off saw　鋸軸移動横切り丸鋸盤
travel device, travel gear, travelling equipment　走行装置

traverse：to travel or lie or extend across ; a lateral movement across something	**移動する，横行する**

This is not a lead screw for screw cutting but for **traversing** the carriage for ordinary turning.

これはネジ切り用の親ネジではなく，普通の旋削用に往復台を**移動させる**ためのものである．

Circuit refers to the path which an electric current **traverses**.

回路とは，電流が**移動する**通路のことである．

The table **traverses** back and forth on accurately machined ways.

テーブルは，正確に機械加工した案内面の上を前後に**動く**．

The workpiece is rotated at high speed and the tool **is traversed** from the workpiece center line rapidly outwards.

工作物は高速で回転し，バイトは工作物の中心から外に向かって迅速に**動く**．

The head and tailstock units were mounted on a **traversing** table. The length of **travel** of the table was automatically controlled by adjusting trips at the front of the machine.

主軸台と心押し台は，1つの**往復動**テーブル上に取付けられていて，テーブルの**移動**長さは，機械前面のトリップ（ひき外し具）を調整することによって自動的にコントロールされる．

At the end of each **traverse**, table stops momentarily.

各横行**運動**の終わりで，テーブルは瞬時に停まる．

〈用　語　例〉

rapid traverse mechanism, quick traverse mechanism　早送り装置
traverse feed　横送り
traverse grinding　トラバース研削
traverse mark　送りマーク
traverse motion device　あやふり装置
traversing table type　テーブル移動形，テーブル往復形
workhead traversing type　工作物主軸台移動形

終わる，してしまう（終了，完了，完結）

complete : to finish (a piece or work etc.)	してしまう，完了する，仕上げる

A final check with a micrometer is made to validate the tool setting, and the cut **is completed**.

ツールのセッティングを確めるために，マイクロメータで最終チェックをして，この切削は**完了する**.

A numerically controlled facing head can **complete** a job like internal threading up to six times faster.

数値制御面削りヘッドは，雌ネジ割りのような仕事を6倍も早く**してしまう**ことができる.

The machine **completes** each part in 16 seconds or 225 parts per hour.

この機械は，各部品を16秒，すなわち1時間に225の部品を**仕上げる**ことができる.

Almost any combination of turning operations can **be completed** in one setup.

どんな組合せの旋削作業も，ほとんど1回の段取りで**してしまう**ことができる.

We previously machined our blank molds in two operations. Now with indexing chucks and G&L designed hydraulics and control software, we **complete** both ends in one operation. . . and with improved quality !

以前はモールド粗材を2回の作業で切削した. 今は割出しチャックとG&Lが設計した油圧装置と制御ソフトウェアで，1回の作業で両方を**やってしまう**…しかも良い品質で.

……. The third is to eject the finished part at **the completion of** all operations.

……．第3は，すべての作業**が完了**した時点で，仕上った部品を突出すことである．

On some machines the work table can perform an automatic cycle of predetermined movements: e.g, a fast run to the cutting position, a change to slow feed motion during the actual cutting, and a quick return to the initial position on **completion** of the cut after which the cycle is repeated.

ある機械では，ワークテーブルがあらかじめ決められた自動サイクルで動く；たとえば，最初に切削位置まで高速で動き，実切削中は送り運動が遅くなり，そして切削**を完了する**と最初の位置に高速で戻る，というサイクルを繰返す．

end : to bring an finish, come to an finish　（完結）する

……put an **end** to the work within half a year.

半年以内に，この仕事を**終了する**．

〈用　語　例〉

end of text character　テキスト終結文字

end of transmission character　伝送終了文字

finish : to bring or come to an end, to complete　してしまう，仕上げる

……, We'll **finish** the job within 30 days, but more often as fast as 10 days.

……．この仕事は30日以内に**終えてしまう**が，大概は10日ですむ．

送　る

feed : to pass a supply of material to　送る，送り量，供給する

In any grinding operation where the wheel **is fed** in a direction normal to the workpiece (infeed), the **feed**, which is the

どんな研削作業でも，砥石は工作物の直角方向に**送られるが**（切り込み），1回の加工ストロ

depth of the layer of material removed during one cutting stroke, will initially be less than the nominal **feed** setting on the machine.

ーク中に除去される材料の深さ，すなわち**送り**は，最初は機械にセットした公称**送り量**よりも小さい．

A cylindrical surface can be generated by **feeding** the tool parallel to the axis of workpiece rotaition.

円筒の表面は，工作物の回転軸に平行にバイトを**送る**ことによってつくることができる．

The tool **is fed** along its axis of rotation.

バイトを，回転軸に沿って**送る**．

Feed the compound **in** 0.005 inch and reset the cross feed dial to zero.

複合刃物台を0.005インチ**送り込んで**，横送りダイヤルをゼロにセットし直す．

Feed the tool **in** 0.002 inch on the compound dial.

バイトを，複合刃物台のダイヤルで0.002インチ**送り込む**．

Usually the work **is fed** against the teeth, the work-feed direction (in relation to the cutter) being longitudinal, transverse or vertical.

通常，ワークは歯に当たる方向に**送られる**．ワークの送り方向（カッタに対して）は，縦，横または上下方向である．

A tapping head **feeds** the tap into the prepared hole using a screw mechanism.

タッピング・ヘッドは，ネジ機構を使って，あらかじめ作ってある穴にタップを**送り込む**．

The wheel **is fed into** the workpiece without traverse motion applied to form a groove.

溝を加工するのに横行運動を与えることなく砥石を工作物に**送り込む**．

On many drill presses, the tool **is fed** by the manual operation of a lever to the right of the head.

多くのボール盤は，ヘッド右手レバーの手動操作によってツールを**送る**．

Hand **feeding** is best at first.

最初は，手**送り**が一番いい．

If this **infeed** of a thread **is made** with the cross slide, equal size chip will be formed on both cutting edge of the tool.

横送り台を使ってネジを**切り込む**場合には，同寸法の切り屑がツールの両方の切れ刃上にできる．

The **feed** (advancing) and adjustment movements of the slides can be performed by means of crank handle on the saddle.

送り台の**送り**（前進）および調整の動きは，サドル上のクランク・ハンドルによって行なうことができる．

Feed is the distance moved by the tool relative to the workpiece in the feed direction for each revolution of the tool or each stroke of the tool or workpiece.

送りは，工具の各回転，あるいは工具または工作物の各ストロークに対し，送り方向に工作物と工具が相対的に動いた距離である．

A roughing **feed** could be from 0.005 to 0.015 inch and a finishing **feed** from 0.003 to 0.005inch for steel.

荒削りの**送り**は，鋼については0.005～0.015インチ，仕上げの**送り**は0.003～0.005インチといったところである．

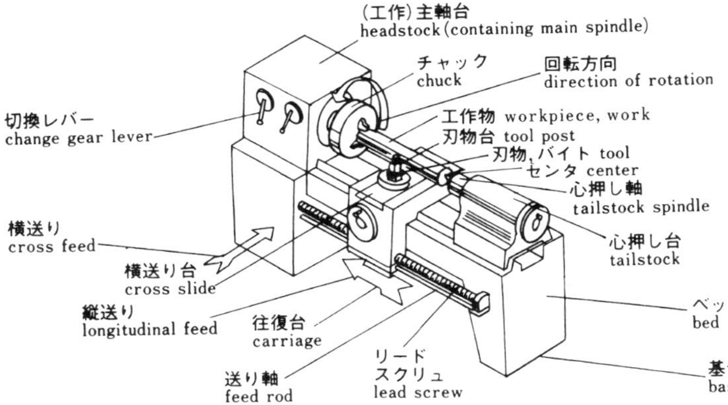

旋盤
lathe

Feed ranges from 0.01 to 0.1 ipr.

送り範囲は，0.01～0.1ipr である．

A suitable **feed** would be 0.35mm, giving an ideal surface roughness of about 3μm arithmetical mean value.

適切な**送り**は0.35mm で，これで算術平均値約 3 μmの理想的な表面粗さが得られる．

Feed is engaged by simple turning the handfeed lever 90° from normal.

普通の位置から90°回すだけで，**送りがかかる．**

Disengage the power **feed**.

機動送りをはずす．

The **feed** should be set at the maximum possible.

送りは，できるだけ最大にセットすること．

The **feed is applied** to the workpiece in increments at the end the return stroke of the ram by a rachet and pawl mechanism driving the leadscrew in the cross rail.

送りは，クロスレールの親ネジを動かす爪車と爪機構によって、ラムの戻り工程の終わりで，一定量ずつ，工作物に**与えられる．**

Finish turning or boring operations are performed at light **feeds** of 0.010 to 0.015 ipr at surface speeds of at least 350 fpm.

仕上げ旋削または中ぐり作業は，少なくとも350fpm の表面速度で，0.010～0.015ipr の小さい**送り量**で行なう．

If the first cut is designed to remove a large amount of material at high **feed** (roughing cut), the forces generated during the operation will probably have caused significant deflections in the machine structure.

最初の切削を，大きな**送り量**で（荒切削）大量の材料を切削するように設計すると，作業中に生じる力によって，多分機械構造がかなり撓むことになろう．

The **feed** in shaping operation was limited to 0.13 mm because of the maximum cutting force the machine could provide.

形削り作業の場合の**送り**は，機械の具備する最大切削力によって0.13mm までに限定された．

The amount of stock **feed** is quickly and accurately set by means of a knurled knob. The length **being fed** is shown on the scale on the **feed** side guide.

材料の**送り**量は，綾目をつけたツマミで，迅速かつ正確にセットされる．その**送られる**長さは，**送り**滑り案内面上の目盛で示される．

Depth of **feed** is determined by adjustable depth-stop screws.

切込みの深さは，調整可能な深さストップネジで決める．

All **feed** controls are located in the pendant.

すべての**送り**制御装置は，吊下がっている．

Automatic control of the **feed motion** may be provided by means of the so-called **feed** shaft, which receive its rotational motion from the work spindle.

送り運動の自動制御は，ワークスピンドルからその回転運動を受ける，いわゆる**送り**軸によって与えられる．

A. has **feed** rate of 0 to 15 ipm on X axis.

Aは，X軸の**送り**速さが0～15 ipmである．

〈用　語　例〉

diagonal feed　斜め送り
feed mark　送りマーク
in feed　インフィード
infeed grinding　送り込み研削（心無し研削盤の）
infeed rate　切込み速度
jump feed　ジャンプ送り
power feed　機動送り
radial feed　半径送り
radial feed screw　切込み送りネジ
sensitive feed　微動送り
skip feed　スキップ送り
through feed　通し送り

traverse: moving the workpiece to another place　　**送る**

The machine has **rapid traverse** of 240 ipm.

この機械は，240ipmの**早送り**である．

High-powered stepping motors provide a **rapid traverse rate** of 200ipm.

強力ステッピングモータで，200ipmの**早送り**ができる．

400ipm **rapid traverse rate**, X and Y

X軸，Y軸の**早送り速度**400

axis, 300ipm **rapid traverse rate**, Z axis.

ipm，Z軸，**早送り速度**300ipm.

Traverse. The V-5 has a 400m/min **rapid traverse** in all axis. Its feed rates range 13 to 5,080mm/min in all axis.

送り. V-5の**早送り**は，全軸400m/min. その送り速度範囲は，全軸13～5,080mm/min.

Upon reaching a specified depth, the drill is **rapidly traversed** back to the original starting position. Then the tool **makes** another rapid **traverse** to the previous preprogrammed depth.

指定の深さに達すると，ドリルは迅速に最初の出発位置に**早戻り**する．つぎに，ツールは，あらかじめプログラムした深さまで，もう一度**早送り**する.

形づくる（成形，整形，細工）

fashion : to make into a particular form or shape

（特定の）形にする

He **made** one of his first violins while a prisoner in a Nazi-camp in Germany, **fashioning** tools from tin cans and gluing scraps of spruce for the upper part.

彼は，ドイツのナチ・キャンプの捕虜になっている間に，缶詰の缶から道具を**つくり**，そして上方の部分にハリモミの廃材を粘付けして，最初のバイオリンの1つを**つくった**.

Increasingly, extruded parts are replacing those formerly **shaped** from tubing by the hot swaging method. These large, hollow components such as wheel spindles and axles **are being fashioned by** a combination of forward and backward extrusion.

押出し部品が，熱間据込み法でチューブから**成形した**ものにとって代わりつつある．輪軸およびアクスルのような，これら大きな中空部品は，前方および後方押出しの組合せで**成形されている**.

form : shape of something ; its outward or visible appearance, to shape ; to mould ; to produce or construct to bring into existence

（形）づくる，形成，成形

Metal, often aluminum, is condensed onto the wafer, filling these gaps and **forming** conducting pathways.

金属（アルミであることが多い）をウェハの上に凝縮してこれらの間隙を埋め，電導性のある通路を**形づくる**.

Developments in tool geometry, in particular, now makes it possible to **form** internal threads in materials harder than 35 Rc and in many different thread **forms.**

特に，ツール形状の進歩によって，今では35Rcよりも硬い材料に，いろいろなネジ**形状**の雌ネジを〈形〉**つくる**ことができる.

When **forming** steel sheet, thickness of 0.045 inch and less an epoxy **form** die can be used, since **forming** pressure will not be excessive.

0.045インチ以下の薄鋼板を**成形する**ときには，**成形**圧力があまり高くないので，エポキシ**成形**型を使うことができる.

Thus a panel can **be** punched and then **formed** on the one machine, using one or two operators as desired.

このように，必要に応じて1～2人の作業者が，1台の機械でパネルを打抜き，ついで**成形する**ことができる.

The machine **forms** parts at 1,200 per hour.

この機械は，毎時1,200個で，部品を**成形する**.

Crank. -- A lever or arm **formed** on a shaft, its object being to convert reciprocating motion into rotary motion.

クランク――軸に**成形した**レバーあるいはアームで，その目的は往復運動を回転運動に変換することである.

Slots **are formed** in the end of the shaft,

スロットは，軸の端部に**つくられている**.

Form a small ball at the end of the glass tube.

ガラス管の端に，小さい球を〈形〉**つくる**.

Form the sample into a cylindrical

サンプルを，円筒形に**つくる**.

shape.

Insert geometry also includes an upward sloping surface behind the depression. The purpose of this surface is to **form** the chips into tight curls for efficient chip disposal.

インサートの形状にもまた，へこみのうしろに上向きの傾斜がある．この表面の目的は，切屑を効率良く処理するために，**切屑を固く巻くように形成する**ことである．

When fitting a new door, some trimming of the door at the edges and some **reforming** with a soft mallet may be necessary to achieve a good fit.

新しいドアを取付けるときには，ぴったり合わせることができるように，ドアの縁をトリミングしたり，軟いハンマーで若干**整形すること**が必要なことがある．

Forming is accomplished in a 500 ton hydraulic press.

500トン油圧プレスで，**成形をする**．

In general, severe cold **forming**, such as drawing and spinning, **is done** with annealed material of the lowest possible hardness.

一般に，引き抜きおよびへら絞りのような強い冷間**成形**は，できるだけ低い硬さの焼鈍材料で**行なわれる**．

mould(mold) : to cause to have a certain shape, to produce by shaping

(形を) つくる，成形，成型，鋳込む

The valves consist of a buna-N packing **moulded** about a steel stem.

弁は，鋼製ステムまわりに**モールドした**ブナNパッキンでできている．

PTFE has not and cannot **be** successfully **molded into** the precise geometric shape required for most spring-energized seal designs.

4弗化エチレンは，バネ付シール設計に必要な，精密な幾何学的形状に**成型された**ことも，またうまくできたこともない．

shape: an area or form with a definite outline : to develop into a certain shape or condition : to give a certain shape to	（形を）つくる

Nozzles **are** specially **shaped** so that the coolant is directed and drawn into the interface between the wheels and blades.

ノズルは，冷却剤をホイールとブレード間の合わせ部分で吸い込むように，特殊な**形につくられている**．

Shape it with a manual pressurized machine under reduced pressure.

手動加圧機械を使い，低い圧力で**成形する**．

気をつける（注意，留意：用心，配慮，考慮）

attention: applying one's mind to something, mental concentration, awareness, consideration	気をつける（配慮）

Where a torque tube is used between a cross chassis member and the rear axle, see that the center joint is rigidly attached to the cross member and **pay** careful **attention** to the lubrication of the joint.

クロスシャーシ・メンバーとリアアクスルの間にトルクチューブが使われている場合には，センター・ジョイントがクロス・メンバーにしっかり取付けられていることを注視し，かつジョイントの潤滑によく**注意を払う**．

Surface damage can be minimized if proper **attention is paid** to the process.

処理に適切な**注意を払う**なら，表面損傷を最小限に留めることができる．

Give special **attention to** the oil pump and sump to make sure that there is no

油ポンプおよび油溜に，確かに漏れがないように，特別に**注**

leakage at these points.

意する.

……unless special **attention is given to** lubricating and designing this cone rib/roller-end contact.

この内輪のツバとコロ端面の潤滑および設計に特別の**注意を払**わない限り，…….

The hands should also **receive attention**, otherwise it may be difficult to get them really clean.

手にも**注意する**こと．そうしないと，本当にきれいにすることはむずかしい．

Like other precision operations, fine-blanking demands more **attention** to details.

他の精密作業同様，精密打抜きは，細かい点により**注意が**必要である．

Apart from an occasional drop of oil, these motors need scarcely any **attention**.

たまたまの油滴は別として，これらモータはほとんど**気をつける必要**はない．

The dynamo is in need of **attention.**

注意を要**する**のはこのダイナモである．

…, and there are usually no more than six or seven points which require **attention** with the grease-gun or oil-can.

普通，グリース・ガンや油差しを使うときに**注意する**ポイントは，6～7ヵ所以下である．

Attention to the piston rings is only necessary at intervals exceeding 30,000 miles.

このピストンリングは，30,000マイル以上の間隔で**気をつければ**よい．

Personnel should be thoroughly trained in handling of fluids and elastomers, with particular **attention** focusing on the proper use of grease and solvents and the elimination of contaminants.

グリースおよび溶剤を適切に使うことと，汚染要因物を無くすことに特に**注意**の焦点を絞って，液およびエラストマの取扱いを完全に訓練すべきである．

…without **attention** on the part of the

ドライバのほうは何も**気を使**

driver.

わないで…….

Performance can usually be improved to a far greater extent by careful **attention** to the efficient working of all moving parts than by any "stunt" tuning.

普通，性能は，どんな高度な技術による調整よりも，動く部品すべてが効率良く働くように**注意することに**よって，ずっと大幅に良くすることができる．

Therefore it is not wise to neglect **attention** to any item that …….

であるから，……のいかなる項目にも**注意**を怠ることはよくない．

The faulty engine should be returned to the manufacturers for **attention**, or replaced by a new one.

欠陥のあるエンジンは，**処置のために**メーカーに戻すか，または新品と交換すること．

care: caution to avoid danger or loss　　**気をつける**

In removing the cylinder-head gasket **care should be taken** that it is not bent or kinked unless it is to be replaced by a new one.

シリンダヘッド・ガスケットを取外すときには，新しいものに交換することになっていない限り，曲げたり，もつれたりしないように**気をつける**．

When refitting **care** must **be taken** to see that ……

ふたたび取付けるときには，……を見るように必ず**注意すること**．

When dual-row thrust bearings are used, more **care is taken** to vent the bearing housing to ensure equal flow through both rows of bearings.

複列スラスト軸受が使われているときには，軸受の両方の列を同じ流量が流れるように軸受ハウジングの穴に一段と**注意する**．

This procedure can be repeated for the

各ピストンについて同じ**注意**

other pistons, **taking** the same **care** with each.

を払って，他のピストンについてもこの手順を繰返すことができる．

For these reasons, therefore **care** should **be exercised** not to breathe an atmosphere contaminated with the exhaust fumes of car engines.

したがって，これらの理由で，車のエンジンの排煙で汚れた大気を吸い込まないように**気をつける**．

Particular **care** should therefore **be paid to** the choice of electrode to be used.

そのため，使用しようとする電極の選択には特別の**注意を払う**．

Even then **care** is needed or the metal may buckle and crack.

そういうときでも**注意**が必要で，そうしないと金属が坐屈し，亀裂することがある．

The synchromesh adjustment requires considerable **care**.

シンクロメッシュの調整には，かなりの**注意**を必要とする．

Great **care** is necessary in tightening these head down evenly, to avoid distrotion.

これらのヘッドを均等に締付ける場合には，変歪しないよう，十分な**注意が**必要である．

If a new fiber gear wheel is to be fitted to the end of the camshaft, this should be pressed on **with** the greatest **care**.

もし，新しいファイバ製歯車を，カム軸端に取付けるときには，最大の**注意を払い**ながら圧入する．

After grinding in, **be** very **careful** to wipe away all traces of the grinding material, both in the valve and in the cylinder.

すり合わせしたあと，弁およびシリンダ両方の研削材のすべての痕跡を拭い去るように，よく**注意する**．

caution : to warn　　**警告する，注意する**

CAUTION: Air will escape from the pipe plug hole while the clutch is set.

注意：クラッチをセットしても，パイプの栓の孔から，エアが逃げる．

Caution must **be exercised** when designing fixtures to be sure that neither they nor the part will flex under high thrust loads.

取付具を設計するときには，高スラスト荷重下でも，取付具，部品のいずれもがたわまないように，**注意し**なければならない．

note: to notice, to pay attention to	**注意する**

…. It should **be noted** that ……

…．**注意す**べきことは……

Publishers' **note**.
American readers of this book should **note** that what is termed a valve in the UK is called a tube in the USA.

発行者**注記**．
この本のアメリカの読者は，英国でいうバルブが，米国ではチューブということに，**注意す**ること．

precaution: something done in advance to avoid a risk	**用心する，予防策**

Some **precautions** should **be taken** when clamping finished surfaces.

仕上表面をクランプ（固定）するときには，何らかの**予防策を講ずる**こと．

The life exɒectancy of these universal joints can be prolonged by using a few **precaution**.

自在継手の寿命は，2～3の**注意**によって，延ばすことができる．

The **precautions** are necessary for operator safety.

オペレータの安全のための**予防処置**が必要である．

chip load: cutting the workpiece with tooth	**切り込む**

chip load of 0.178mm/tooth

1歯当り，0.178mmの**切込み**

Chip loads per tooth (depths of cut) exceeding about 0.010″ (0.25mm) on 319-T5 alloy can cause breakouts on the workpiece edges of more than 1/16″ (1.6 mm).

319-T5合金に，1歯当りの**切込み**（切削の深さ）が，約0.010インチを超えると，工作物の縁が1/16インチ以上欠けることがある．

A new K. cutting tool material can cut cast irons and nickel base alloys at higher cutting speeds with **heavier chip loads** than either solid ceramics or ceramic-coated carbides.

新しいK.バイト材は，ムクのセラミックスやセラミック被覆の超硬よりも高切削速度，**重切込み**で鋳鉄およびニッケル合金を切削できる．

〈用　語　例〉

cross slide　切込み台
depth setting〔plunge grinding〕　切込み量
entering angle, entrance angle　切込み角
feed cam　切込みカム
radial feed screw　切込み送りネジ

cut into:	**切り込む**

Cut into core as little as possible.

コアへの**切込み**は，できるだけ少な目に．

depth of cut:	**切り込み**

Feed and **depth of cut** should be as much as the tool, workpiece, or machine can stand without undue stress.

送りおよび**切込み**は，バイト，工作物，機械が過大な応力なく，耐えうる程度にする．

engage: to interlock (parts of gear) so that it transmit power; to become interlocked in this may	**切り込む，噛合う**

It also can **engage** on both inside and outside surfaces.

また，内・外両面に**切り込む**ことができる．

The **back engagement** (previously known as depth of cut) be kept small, about equal to the feed.

噛合い（従前，切込み）は小さく，送りとほぼ等しくする．

In general the **back engagement** determines the depth of metal removed from the workpiece in a single-point cutting operation.

一般に，**切込み**（バック・エンゲージメント）で，バイト切削作業における工作物から取り去られる金属の深さが決まる．

In slab milling the **working engagement** was known as depth of cut.

スラブ・フライス削りにおける**ワーキング・エンゲージメント**とは，これまでの切込みのことである．

The wheel-workpiece **engagement** (width of grinding path and feed) will change as cutting proceeds.

砥石と工作物の**噛合い**（研削パスの幅および送り）は，切削が進むにつれて変わる．

From Fig. 1．6, therefore, a_c is given by $a_f \sin K_r$, where a_f is the **feed engagement**, the instantaneous **engagement** of the tool cutting edge with the workpiece measured in the direction of feed motion.

したがって，図1．6から，a_c は $a_f \sin K_r$ で与えられる．ここに，a_f は**送りのエンゲージメント**で，送り運動方向で測ったバイト切れ刃と工作物の瞬間的**噛合い**である．

切る，削る，切削する

broaching；	broach : to make a hole in and draw out liquid.	**ブローチ削り**

Broach a ¼ inch keyway in a 1¼ inch bore.

1¼インチの穴に¼インチのキー溝を**ブローチで削る**．

Broaching is more complex: Parts positioned with precision locators, two sides of each part **are** straddle **broached**, locators are retracted, ends of parts **are broached** flat, plus two chamfers each. 880 parts per hour.

ブローチ削りは，大変複雑である．部品（複数）を精密位置決め具で位置決めし，各部品の両側をまたぐようにして**ブローチで削る**．位置決め具が引っ込むと，部品の各端の平面と2つの面取りを1時間当り880個，**ブローチで削る**．

Broaching is mainly employed for machining out holes or other internal surfaces,……. The cut starts with the smaller teeth, which enter the hole, and finishes with larger teeth, which bring the hole to the finished size.

ブローチ削りは，主として穴などの内面を切削するために使われるが……．切削は，穴に入る小さいほうの歯で始まり，穴を仕上寸法にする大きいほうの歯で終わる．

The **broaching** operation is performed by a machine that pulls or pushes the broach through the workpiece.

ブローチ削り作業は，工作物を貫通してブローチを引く，あるいは押す機械で行なわれる．

〈用　語　例〉

broach sharpener　工具研削盤，ブローチ研削盤
internal broaching machine　内面ブローチ盤
pull broaching　引きブローチ削り
push broaching　押しブローチ削り
surface broachig machine　表面ブローチ盤

contour: an outline　**輪郭削り**

The Series MVC can be used for simultaneous three-axis continuous path linear **contouring** with circular **contouring** in any plane.

MVCシリーズは，どんな面の同時3軸連続輪郭**削り**（あらゆる面の円弧**削り**を含む）にも使うことができる．

〈用　語　例〉

contour grinding machine　輪郭研削盤
contour milling machine　輪郭フライス盤
contour sawing machine　金切り帯鋸盤

cut: to divide or wound or detach with an edged instrument, to shape or make in this way.	**切削する，機械加工する，削る，切る，カットする，切断する**

After a few full threads **are cut**, the die should be removed so that the thread can be tested with a nut or thread ring gauge.

ネジを二，三山完全に**切ったら**，そのネジをナットあるいはリングネジ・ゲージでテストできるように，ダイスを取外す．

The method is especially suitable for soft metals, such as aluminum, which are difficult to screw-**cut** with a smooth finish.

この方法は，滑らかな仕上げのネジ**切り**はむずかしい，アルミニウムのような軟かい金属には特に適している．

A groove of this type should **be cut** with a round nose tool.

この種の溝は，丸先バイトで**切削する**．

A typical relationship between rake angle and tool life is shown in Fig.4.7, where the optimum rake is approximately 14° when **cutting** high strength steel with a high-speed steel tool.

すくい角とバイト寿命の代表的な関係を図4．7に示す．図で，高強度鋼を高速度鋼バイトで**切削する**とき，その最適すくい角は約14°である．

A long screw thread of coarse pitch can suitably **be cut** by means of a disc-shaped single cutter.

粗いピッチ（並目）の長ネジは，ディスク形シングルカッタで，うまく**切る**ことができる．

Cut Acme threads on the lathe.

旋盤で，アクメネジを**切る**．

Any given shape, punch and die, cam or gear rachet, in fact, anything that can **be cut** by Agietron's DEM 15/30 machines, can be programmed with the LAMA 25 system in the shortest possible time.

与えられたどんな形状のパンチやダイ，およびカムやラチェットなど，実際にアジエトロンのDEM15機で**加工**できるものなら何でも，最短時間でLAMA25システムでプログラムできる．

Screw threads, both external and internal, can also **be cut** efficiently and economically by milling. This is done on a milling machine, so-called thread milling cutter being used for the purpose.

ネジ（雄ネジおよび雌ネジ）は，フライス削りで，効率良くかつ経済的に**切る**ことができる．これはフライス盤で行ない，これ用のいわゆるネジフライス・カッタが使われる．

The gearing enables screws of varying pitch and diameter to **be cut** by varying the speed of rotation of the lead screw.

この歯車装置は，親ネジの回転速度を変えることによって，ピッチおよび直径の違うネジを**切ること**ができる．

The tool is rotated by hand, the workpiece being gripped in a vise. Alternatively, the workpiece be gripped in the chuck of a lathe and rotated, while the tap or die is guided by the tailstock. The screw tap **cuts** an internal thread in a hole drilled beforehand. The screw die **cuts** an external thread on a rod or bolt and is held in a device called a stock, which has handles for manipulation.

切削工具を手で回し，工作物をバイスでつかむ．代わりに工作物を旋盤のチャックにくわえて回してもよいが，タップやダイスは心押台でガイドすることになる．（ネジ）タップでは，前もって穴明けした穴に雌ネジを**切る**．ネジダイスはロッドやボルトに雄ネジを**切る**が，操作用のハンドルの付いたストックという道具で保持する．

H. first **cut** his gear slightly oversize on his gear shaper. Then, after hardening, it was finished to exact size on his gear-cutter grinding machine.

Hはまず，（彼の）歯車形削り盤で若干オーバーサイズに歯車を**切って**，つぎに焼入れしてから，（それを彼の）歯切り研削盤で正確な寸法に仕上げた．

Bores of felts should **be** clearly **cut** to the diameter of the shaft.

フェルトの内径は，軸の直径にきれいに**切る**．

The fact that the T-Max U carbide drill

Sandvik 社（Fair Lawn, NJ）

from Sandvik, Inc., Fair Lawn, NJ, **cuts** to size from solid means that cycle time is **cut** in half.

のT-MAX U超硬ドリルで所定寸法にムクのものから**切削することは**，サイクルタイムが半分に**削減**されることでもある．

Machine a nut blank and **cut** a 1-8 UNC internal thread to fit a plug gauge.

ナットのブランクを機械加工し，プラグ・ゲージに（ぴったり）合うように1-8 UNCの雌ネジを**切る**．

He also made planing machine which could **cut** in both direction of the table.

彼はまた，テーブルの両方向で**切削**できる平削り盤を製作した．

The right-hand tool has the cutting edge on the left side and **cut** to the left or toward the head stock.

右曲りバイトの切れ刃は左側で，左のほうすなわち主軸台の方向に**切削**する．

When **cuts are made** at any angle other than right angle to the horizontal or the

水平あるいは垂直に対して直角以外の角度で**切削する**とき，

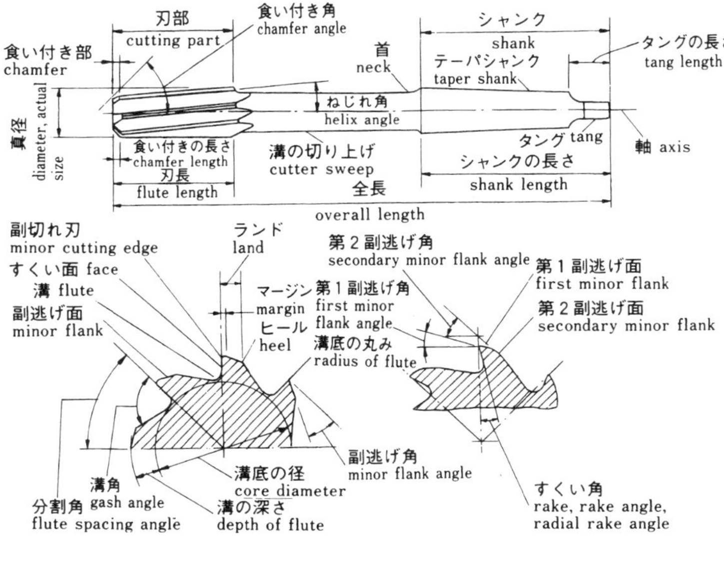

vertical they are called angular cuts.

それをアンギュラー・カットという.

Tools that have special shaped **cutting edges** are called formed tools. These tools are plunged directly into the work, **making the full cut** in one operation.

特殊な**切れ刃**形状のバイトを，総形バイトという. これらのバイトは，ワークにプランジ切込みし，1回で**完全に切削する**.

Magnetic chucks are sometimes used for **making** light **cuts** on ferromagnetic materials.

マグネチック・チャックは，時に常磁性材を軽**切削する**ために使われる.

Thread cutting on a lath with a single point tool is done by **taking** a series of **cuts** in a helix of the thread.

旋盤によるバイトを使ったネジ切りは，ネジのねじれ部に一連の**切削をする**ことによって行なわれる.

A light **cut is taken** along the length of the test-piece and both ends are measured with a micrometer.

テストピースを長手方向に沿って軽く**削って**，両端をマイクロメータで測る.

The lands on these reamers are ground cylindrically without radial relief, and all **cutting is done** on the end of the reamer.

このリーマのランド(当り部)は，(半径方向)逃げなしで円筒に研削する. そして**切削は**すべてリーマの端部で**行なわれる**.

Most metal **cutting is** still **performed** with tools manufactured from high-speed steel.

金属**切削**のほとんどが，今でも高速度鋼製のバイトで**行なわれている**.

As soon as the **cut is completed**, the tilt arm quick returns to its starting position.

切削がすむやいなや，ティルト・アーム（傾斜腕）は初めの位置に迅速に戻る.

A higher degree of precision can be obtained with machine-**cut** screw threads,

（機械）**切削**ネジ，特に旋盤で作ったものは，精度が高い.

especially those produced on a lathe.

The diameter of the workpiece is checked with a micrometer and the remaining amount **to be cut** is dialed on the cross feed micrometer.

工作物の直径をマイクロメータでチェックし，横送りマイクロメータの目盛を**切削すべき**残量に合わせる．

the part left **uncut**……

切削しないで残った部分は……

〈用　語　例〉

balance cutting　釣合い削り
bevel cut　ベベル切断
combined cutting　複合削り
cross-cut band saw machine　横切帯鋸盤
cut bolt　切削ボルト
cut-off　突切り
cut〔thread〕tap　切削タップ
cut thread　切削ネジ
cutting angle　切削角（削り角）
cutting fluid　切削油剤，加工油剤
cutting oil　切削油
cutting speed　切削速度（削り速度）
cutting stroke　切削工程
cutting tool　切削工具
draw cutting　引削り
finish cutting　仕上げ切削
gear cutting　歯切り
gear cutting machine　歯切り盤
internal gear cutting attachment　内歯車削り装置
lead screw cutting　親ネジ切り
lead screw cutting lathe　親ネジ切り旋盤
multitool cutting　多刃削り
plunge-cut mechanism　プランジカット機構
rack cutting machine, rack shaper　ラック歯切り盤
rough cutting　荒削り
screw cutting lathe　ネジ切り旋盤
thread cutting　ネジ切り
thread cutting device　ネジ切り装置
thread cutting lathe　ネジ切り旋盤
under cut　切下げ，アンダーカット

facing : to shave a right face against an axis

面削り

Correctly set up a workpiece and **face** the ends.

工作物を正しくセットして，端を**面削り**する．

After heat treating to RC62～64, the ends of the bushings **are faced** so that identification markings can be read. Following

RC62～64に熱処理後，ブシュの端面を識別マークが読めるように**面削りする**．外径の心無し

centerless grinding of the O.D. the bushings are chucked and then ground on the ID. Certain sizes and models of bushings are lapped and honed.

研削に続いて，ブシュをチャックでつかみ内径を研削する．特定の寸法および形式のブシュについては，ラップ仕上げおよびホーニング仕上げする．

Facing to length may **be accomplished** by trying a cut and measuring with a hook rule or by **facing** to a previously made layout line.

試しに１回削って鉤形尺で測るか，あるいはすでにけがいておいた線まで**面削りする**ことによって，所定の長さに**面削りすることができる**．

A facing operation can be carried out by using a special toolholder (Fig. 1.15) that feeds the tool radially as it rotates.

面削り作業は，回転するにつれて，バイトを半径方向に送る特殊ツールホルダ（図1.15）を使うことによって行なうことができる．

groove: a long narrow channel in the surface of hard material
grooving: to make a groove or grooves in

溝削り

Special tools (Figure 38) are ground for both external and internal grooves and recesses. Parting tools are sometimes used for external **grooving** and thread relief.

特殊なバイト（図38）は，外面および内面の溝および凹み用に研削されたものである．突切りバイトは，時に外面の**溝削り**およびネジの逃げ切りに使われる．

knurling; knurl: bead or ridge in metal work

ローレット切り

Knurls do not cut, but displace the metal with high pressure.

ローレットは切削するのではなく，高圧力で金属を押し除ける．

machine : to make or produce on (a thing) with a tool

切削する，機械加工する（加工する），削る

One pass of a giant broach equipped with teeth of tungsten carbide can finish **machine** the face of a cylinder head or cylinder block.

超硬歯を付けた巨大なブローチの１回通しで，シリンダヘッドや，シリンダブロック面を仕上げ**加工**できる．

Component **is** forged and finish **machined**.

部品は鍛造して仕上げ**切削する**．

When a large batch of components are **to be machined**,……

大ロットの部品を**機械加工することになっている**ときは……

Machine components from bar stocks (work material in bar form).

加工部品は棒材（棒状の物）から（機械加工で）作る．

A live center **is** often **machined** from a short piece of soft steel mounted in a chuck.

ライブ・センタは，チャックに取付けた短かい軟鋼の部品で**作られる**（切削する）ことがよくある．

While an average of 10 pieces **were machined** by the cutter coated with A‐2, ……

A‐2を塗布したカッタにより，平均10個まで**切削されるが**，……

Surface finish of the part **machined** with coated and uncoated cutters were measured using a portable surface indicator.

（固体潤滑剤を）被覆したカッタと被覆しないカッタで**切削した**部品の表面仕上げを，ポータブル表面検査機を使って測った．

Figure 5 indicates the magnitude of the lubricant effect when copper **is machined** with a variety of compounds and two

図５は，いろいろな合成剤および塩化物添加剤を含有する２つの市販の（切削）液で（を使

commercial fluids containing chlorinated additives.

って)，銅を**切削した**ときの潤滑剤の効果の大きさを示す．

Specimens **were machined** with machining parameters shown on Table 2. Combinations of feed, speed and radial depth were selected which would yield a tool life of 30 to 90 minutes.

標本を表2の切削パラメータで**切削した**．30～90分のバイト寿命となるような送り，速度，半径方向の切込みの組合せを選んだ．

Live center **is machined** in a four-jaw chuck. It is then left in place and the workpiece is mounted between it and the tailstock center.

4つ爪チャックでライブ・センタを**切削する**．つぎにそれを，そこにそのままにしておいて，工作物をそれと心押台センタの間に取付ける．

With this new tool steel, John Fowler & Co. turned iron shaft in the lathe at the rate of 75 feet per minute, and when **machining** steel wheels in their boring mill they could make roughing cuts 1/2 inch deep.

この新しい工具鋼で，ジョン・フォウラー社は旋盤で鉄のシャフトを75fpmの速度で旋削し，そして中ぐり盤で鋼の車輪を**切削する場合**には，(切込み)深さ1/2インチの荒削りをすることができた．

This is why shafts that are to be frequently finish-ground between centers must **be machined** between centers on a lathe.

これが，センタ支持で頻繁に仕上げ研削するような軸を，なぜ旋盤のセンタ支持で**切削**しなければならないかという理由である．

Parts which can **be machined** on the Super 10 range from high precision aluminum components to steel configurations requiring a multiplicity of precision operations.

このスーパー10で**切削**できる部品(の範囲)は，高精度のアルミ部品から，精密作業を数多く必要とする形状の鋼にまで及ぶ．

The ways of the lathe **are** very accurately **machined** by grinding or by milling and hand scraping.

旋盤の案内面は，研削またはフライス削りと手によるキサゲ仕上げにより，非常に高精度に**加工されている**.

Outer ring seatings **are machined** in two separate operations.

（軸受の）外輪の座は，２つの別々の作業で**機械加工される**.

The above materials **are rough machined** either in the annealed state or in the normalized condition.

上記材料は，焼戻し状態あるいは焼ならし状態のいずれかで**荒削りする**.

A gear hardened to 57 H_RC with 25mm module, 250mm face width, 49teeth, and 19° helix angle **was machined** at a speed of 97 m/min and feed of 1.5mm. The hob showed a flank wear of 0.27mm after **machining** seven gears.

モジュール25mm，歯幅250 mm，歯数29，ねじれ角19°の57 H_RC に焼入れした歯車を，速度97m/min，送り1.5mmで**切削した**．このホブは，ギア７個を**切削**したあとの逃げ面の摩耗が，0.27mmであった.

For short runs or production runs, the low cost M. has the capacity to **machine** parts at efficiencies normally associated with machining centers in a much higher price range.

この低コストのM.は，小ロット，あるいは生産加工において，高価格なマシニングセンタが普通もっている効率で，部品を**加工する**能力がある.

A cutter **machines** a shaped slot in the work piece.

カッタで工作物にある形状の溝を**切削する**.

Machine flat surface on small components.

小さい部品に平面を**切削加工する**.

The polyimide is purchased in rod form and the bearing manufacturers **machine** the rods into retainers.

このポリイミド（材）は長棒で購入し，軸受メーカーがこの長棒を（軸受）保持器に**加工する**.

Stub mandrels are also used in chucking operations. These **are** often quickly **machined** for a single job and then discarded.

ずんぐりしたマンドレルは，チャッキング作業にも使われる．これらは，しばしば単一仕事用に手早く**機械加工され**，そして捨てられる．

Machining the bottom or the end of the blind hole to a flat is easier when the drilled center does not need to be cleared up.

穴明けしたセンタをきれいにとる必要がなければ，盲孔の底（端）を平らに**削る**のはより容易である．

The edges **have been machined** true and square to each other.

縁を，正しくかつ互いに直角に**削った**．

machine a bore to a taper of about 5°.

穴を，約5°のテーパに**機械加工する**．

The molded rods or tubes **are** then **machined** to the precise dimensions on the engineering drawings. A great deal of expertise is required to **machine** a plastic in order to achieve surface finish as well as engineering dimensional tolerances.

つぎに，成形した長棒または管を，設計製作図の精密な寸法に**切削する**．表面仕上げならびに設計製作寸法の公差にするためには，プラスチックを**切削する**のに多くの専門的技術が必要である．

For this reason, it is usual to **machine** the workpiece oversize deliberately during the roughing cut, leaving a small amount of material that will subsequently be removed during the finishing.

こういう訳で，荒切削では計画的に工作物をオーバーサイズに**加工し**，つぎの仕上げで削り取る少量の材料を残しておくのが，普通である．

A massive forging **is machined** on a lathe to exact specifications.

嵩ばって重い鍛造品を，旋盤で，仕様通りに**切削する**．

……can **machine** single piece part up to 12″ dia.

……は，単品なら，直径12インチまで**切削**できる．

B. hydraulically actuated multi-position tables are equipped with 2-or 6-position adjustable stops. This permits you to **machine** up to 6 holes in a line.	B油圧作動マルチポジション・テーブルには2または6位置可調整ストッパが付いている．これで，1列穴6個まで**加工**できる．
Machining shoulder to specific length **could be done** in several ways.	いく通りもの方法で，ショルダーを仕様の長さに**切削できる．**
All **machining is done** by the lower end of the tap.	**切削**はすべて，タップの下端で**行なわれる．**
Most production **machining is done** with carbide tools.	ほとんどの生産（機械）**加工**は，超硬工具で**行なわれる．**
Machining with a single point tool **can be done** anywhere on the workpiece except near or at the location of the lathe dog.	回し金のところやその近くを除けば，工作物のどんなところでも，バイトで**切削できる．**
Finish **machining is performed** after quenching and tempering to a high hardness value.	高硬度に焼入れ，焼戻しをしてから，仕上げ**加工する．**
ID machining on a hub for a food industry packaging machine **is performed** on a Warner & Swazey 2AC chucker at the A. Company in Los Angeles.	ロスアンゼルスのA社では，W & S 2ACチャッカで，食品産業用包装機ハブの内径**加工**をしている．
Some believe that by the end of this decade at least one-half of all steel **machining** will **be carried out** using cemented titanium carbide.	この10年代の終わりまでには，鋼**切削**の少なくとも半分が，セメンテッド・チタニウムカーバイドを使って**実施される**だろうと，考えている人もいる．

Machining can **be facilitated** by the following methods: ……

つぎの方法で，**より楽に切削することができる．**

Machining the tiles **is accomplished** with computer controlled milling machines. After their upper surfaces **are** individually **machined**, coated, and water-proofed, the tiles……

タイルはコンピュータ制御のフライス盤で**切削する．**タイルの上面を１つ１つ**切削，**塗装，防水加工してから，……

We also **machine** our parts to extremely fine tolerance:

また，われわれは部品をきわめて高い公差に**切削加工する．**

All **machining** of that family of components **can be performed** in one small area of the machine shop, perhaps with only one or two operators tending the four machines.

この種の部品のすべての**機械加工**は，機械４台に作業者１～２名くらいでよく，工場(内)の小スペースで可能である．

〈用　語　例〉

abrasive machining　砥粒加工
machined bolt　切削ボルト
machined nut　切削ナット
machining ratio　加工比
machined thread　切削ネジ
machining　切削，機械加工
surface to be machined　被削面

milling ; mill: to cut or shape (metal) with a rotating tool　**フライス削り**

Mill flat surface to size.

所定の寸法に，平らな表面を**フライスで削る．**

, and then **mills** the slots in the block.

つぎに，ブロックに溝を**フライスで削る．**

……. Their versatile end mills also slot and pocket **mill**, counterbore, spot-face, ramp and peripheral **mill** on a wide variety of materials.

……．また，この（メーカーの）汎用エンドミルは，広汎な種類の材料に，溝およびポケット**フライス削り**，深座ぐり，座ぐり，斜面および円周**フライス削り**ができる．

The feed (longitudinal motion) of the carriage and the rotation of the workpiece are interlinked by means of a lead screw and gearing so as to obtain the correct pitch of the thread **being milled**.

往復台の送り（縦方向運動）と工作物の回転は，**フライス削り**で正しいネジピッチが得られるように，親ネジおよび歯車（系）で連結されている．

Milling is a machining operation in which a workpiece is given the desired shape by the action of the rotating cutter while the workpiece performs linear movements.

フライス削りは，工作物が直線運動をする間に，回転するフライスの働きによって工作物に所望の形状を与える切削作業である．

Screw threads, both external and internal can also be made efficiently and economically by **milling**. This is done on a milling machine, socalled thread-milling cutters being used for the purpose.

ネジ（雄ネジ，雌ネジとも）は，**フライス削り**で，効率良くかつ経済的に切ることができる．これはフライス盤で行ない，フライス削り用のいわゆるネジフライス・カッタが使われる．

〈用　語　例〉

circular milling　回しフライス削り
climb milling　上向き削り
copy milling　ならいフライス削り
downward milling　下向き削り
end milling　エンドミル削り
face milling　正面削り
form milling　総形フライス削り
gang cutter milling　組合せフライス削り
helical milling attachment　へリカルフライス削り装置
mill　製造工場，ひく
mill finish　圧延仕上げ
miller　フライス盤
milli-　1/1000
milling　フライス削り，フライス切削，フライス作業
milling attachment　フライス装置
milling　フライス，フライス刃物
milling machine　フライス盤
production miller / production milling machine　生産フライス盤
peripherical milling　円周削り
plain milling　平フライス削り
profile milling machine　ならいフライス盤
profiling milling machine　ならいフライス盤
rack milling attachment　ラック切りアタッチメント
side milling　側フライス削り
two-high rolling mill　2段圧延機
upward milling　上向き削り

necking : to make a neck-down　　**逃げ溝削り**

Four different diameters were turned with a **neck-down** section between each diameter.

4つの異なった直径を，各直径の間に**逃げ溝**（部）をつけて旋削した．

profile : a side view, especially of the human face　　**ならい削り，縦断面図**

The part is initially hogged out on a tracer milling machine and **profiled** on a numerical controlled five axis milling machine.

部品は初めにトレーサ・フライス盤でほじくり出し，NC 5軸フライス盤で**ならい削りする**．

〈用　語　例〉

profile grinder	ならい研削盤	profile planer	模様削りカンナ盤
profile milling	ならいフライス削り		

recessing : to make a recess in or of (a wall etc.)

recess : a part or space set back from the line of a wall etc., a small hollow place inside something　　**逃げ溝切り**

Recessing and grooving on external diameters is done to provide grooves for thread relief, snap rings and O-rings (Fig. 37a to c).

外径面の**逃げ溝切り**および溝切りは，ネジの逃げ，止め輪およびOリング用の溝を設けるために行なわれる（図37 a ～ c）．

〈用　語　例〉

recess of roller　　コロのぬすみ

remove : to take off from　　**取り去る，切削する，研削する，加工する**

This reamer will **remove** a considerable

このリーマは，1回の切削で，

amount of material in one cut.

かなりの量を**切削する**.

In Norton lab tests, the standard Norton SD diamond wheel **removed** 135 cu. in. of carbide per cu. in. of wheel wear, where the new wheel **removes** 316 cu. in. of carbide per cu. in. of wheel wear.

ノルトン研究所のテストで，標準のノルトンSDダイヤモンド砥石は砥石摩耗1 $inch^3$ 当り超硬を135$inch^3$ を**研削したが**，新しい砥石では316$inch^3$ **研削できた**.

If more than the tips of the tool marks **are removed** with a file or abrasive cloth, a wavy surface will result. For most purposes 0.002inch is sufficient material to leave for finishing.

ヤスリや研摩布で，バイト目の先以上**削る**と，波状の表面になる．大部分の用途には，仕上げ用に残す材料は0.002インチで十分である.

In ordinary turning, metal **is removed** from a rotating workpiece with a single point tool.

普通の旋削では，バイトを使って金属を回転する工作物から**削りとる**.

Material **being removed** by one cutting edge…

1つの切れ刃で**削られる**材料は…

On a lathe, metal **is removed** from a workpiece by turning it against a single point cutting tool.

旋盤では，工作物をバイトに当てて回転させることによって，金属を工作物から**削り取る**.

The workpiece is measured with a micrometer and the desired length is substracted from the measurement ; the remainder is the amount you should **remove** by facing.

工作物をマイクロメータで測って，測定値から所要の長さを引く；その残りが，面削りで**切削する**量である.

Volume of metal **removed** electrolytically per unit time…

電気分解で**取り去られる**単位時間当りの金属の容積は…

Metal **removal** in ECM is not achieved

ECMの金属**加工**は，機械的

by mechanical shearing (as in conventional machining) or by melting and vaporization of the metal (as in EDM).

剪断（普通の機械加工におけるような）や，金属の溶融と蒸発（EDMにおけるような）でするものではない．

〈用　語　例〉

residual stock removal　切り残し　theoretical stock removal　設定研削量

rough : to shape or plan or sketch roughly　**荒削り**

Rough the side relief angle and the side cutting edge angle.

第一横逃げ角と横切れ刃角を**荒研削する．**

Using the negative rake principle, VR/Wesson came out with very fine pitch cutter with solid carbide blade in 1948 (the

VR/Wessonが，鉄のエンジン部品の高速・高送り**荒削り**用に，ムクの超硬刃付きでピッチ

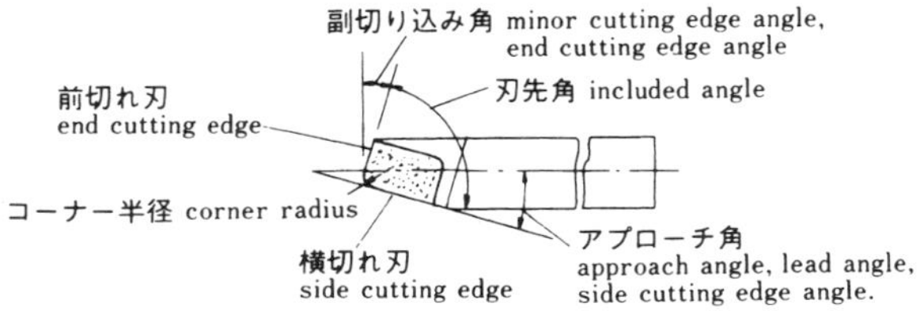

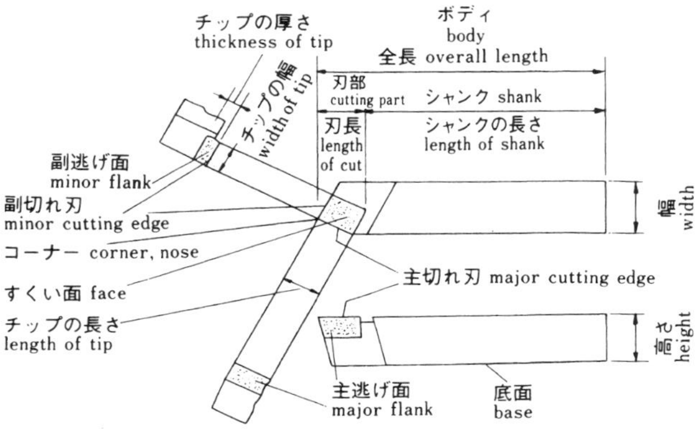

Rigid-Cut 600 series) for high-speed-high-feed **roughing** of iron engine parts.

のきわめて細かいマイナスすくい角法を用いた，フライスを発表したのは，1948年である．

The two end knobs **were roughed out** on a lathe using several angular cuts, then filed to match a radius gage and polished.

両端のノブは，旋盤を使い数回アンギュラ・カットで**荒削り**してから，半径ゲージに合うようにヤスリで仕上げ，そして磨く．

rough out blank to length.

ブランクを適当な長さに**荒取りする．**

And, with the ability of modern casting and forging to produce a part much closer to the final measurements desired, the preliminary **rough-out** is often eliminated and the components can be finish ground from the blank.

鋳造や鍛造で所要最終寸法に非常に近い部品を作ることができるため，**荒削り**を省けることが少なくなく，部品をブランクから仕上げ研削することができる．

〈用　語　例〉

first roughing〔hand〕tap　一番タップ
gear roughing　荒歯切り
rough cut file　大荒目ヤスリ
rough forging　荒地
rough mechining　荒削り
rough roll　荒引きロール
roughing tool　荒削りバイト
second roughing〔hand〕tap　二番タップ

screw : to cut a thread on　**ネジを切る**

Bolt : A metal rod or shaft having a head at one end, the other end **being screwed** to receive a nut or to fit a tapped hole.

ボルト：一端に頭のある金属の棒または長棒で，他端はナットを付けたり，ネジ立てした穴に嵌まるよう**ネジが切られている．**

〈用　語　例〉

automatic screw machine　自動ネジ切り盤
screw cutting　ネジ切り

screw cutting lathe　ネジ切り旋盤
screw-driven planer　ネジ式形削り盤
screw-gear　ネジ歯車
screw-head　ネジ頭

shaping: to shape; to give a certain shape to　**形削り**

The flanges of the valve body **are shaped** in a turning operation at about 470 rpm.
バルブ本体のフランジは，約470rpm の旋削作業で**形を作る.**

〈用　語　例〉
gear shaper, gear planer　歯車形削り盤
rack shaper　ラック歯切り盤
vertical shaper　立削り盤

〈関連用語〉
gear stocking　荒歯切り
generating rolling circle　歯切りピッチ円
rack cutting machine　ラック歯切り盤
round bar making machine　自動丸棒削り盤
slotting machine, slotter　立削り盤
spiral bevel gear generator　マガリバ傘歯車歯切り盤
straight bevel gear generator　スグバ傘歯車歯切り盤
wood wool making machine　木毛削り盤

surfacing: surface; to come or bring to the surface　**サーフェーシング，平面を削る**

Surface a workpiece in a shaper.
形削り盤で工作物の**平面を削る.**

〈用　語　例〉
surface grinder　平面研削盤
tread surfacing moulder　木履表反り形削り盤

thread: to cut a thread on (a screw)　**ネジを切る**

A tap that **threads** over 1000 holes per tap is the result of a change in geometry, a redesign, radical change in surface treatment, metallurgical change, and
1,000以上のが**ネジ切り**できるタップは，寸法，形状の変更，設計のし直し，表面処理の抜本的変更，冶金上の変更，熱処理

revised heat treating specifications.

仕様の改訂の成果である．

The handle was made from a rod on which both ends **were threaded** for end knobs.

ハンドルは棒からつくり，その両端に握り用の**ネジを切った**．

The lead screw is a long **threaded** rod extending along the lathe and serving as a master screw for cutting screw thread.

親ネジは，ネジ切り用マスターネジとして役立つように，旋盤に沿って伸びた**ネジを切ってある**長い棒のことである．

A tap is used for internal **threading**.

タップは，雌**ネジ切り**に使われる．

〈用　語　例〉

circular threading tool　雄ネジ切り丸バイト
external threading tool　雄ネジ切りバイト
inside single-point thread tool　雌ネジ切り〔一山〕バイト
internal threading tool　雌ネジ切りバイト
outside single-point thread tool　雄ネジ切り〔一山〕バイト
thread tool setting microscope　ネジ切りバイト顕微鏡

turning ; turn : to shape in a lathe　**旋削する，外丸削り**

Trepan is to **turn** annular grooves or recesses in, as a block is held in a lathe.

心残し削りとは，旋盤に保持したブロックに，環状の溝やへこみを**旋削**することである．

With this new tool steel, John Fowler & Co., steam-plough builder of Leeds, **turned** iron shaft in the lathe at the rate of 75 feet per minute, and when machining steel wheels, they……

この新工具鋼で，J. Fowler社（Leedsの蒸気プラウ・メーカー）は，旋盤の送り速度75fpmで鉄の軸を**旋削し**，また鋼の車輪を切削するときに……

We **turn** a 50″× 6″ OD shaft of 4343 annealed steel in 1.19 hours now. It used to take 2.14 hours on our old NC lathe.

焼鈍した4343鋼の50インチ×6インチ外径の軸を，今は1.19時間で**旋削している**．古いNC旋盤では，2.14時間かかっていた．

For accurate **turning** operations or in cases where the work surface is not truly cylindrical, the workpiece **can be turned** between centers. This form of work holding is illustrated in Fig. 1.9.	精密な**旋削**作業，あるいはワーク表面が真に円筒でない場合には，ワークをセンタ支持で**旋削する**ことができる．このワーク保持方式を，図1．9に示す．
When a large batch of steel shafts are to **be rough-turned** to 76mm diameter for 300mm of their length at a feed of 0.25mm ……	大ロットの鋼の軸を，送り0.25mm，長さ300mm，直径76mmに**荒旋削する**ようなときには，……
The shaft **is turned down** to form a shoulder.	軸を**削り落して**ショルダーを作る．
To **turn** a workpiece between centers, it is supported between the dead center (tailstock center) and the live center in the spindle nose.	センタ支持で工作物を**旋削する**には，デッド・センタ（心押し台センタ）と主軸端のライブ・センタの間に支持する．
…… involves **turning** diameter and cutting thread with the lathe.	……には，旋盤による外径の**旋削**とネジ切り（作業）もある．
The first completely automatic turret lathe for **turning** metal screw was designed and built by Christopher Miner Spencer.	金属ネジ**旋削**用の完全自動タレット旋盤を，最初に設計・製作したのはC. M. Spencerである．
A more modern method of **turning** to size predictably uses the compound and cross feed micrometer collars and the micrometer calipers for measurement.	想定寸法に**旋削する**もっと新しい方法では，複式刃物台と横送り台のマイクロメータ・カラーおよび測定用マイクロメータが使われている．
A considerable amount of straight **turning** on shaft is done with the work held between a chuck and the tailstock	軸のストレート**旋削**の多くは，工作物をチャックと心押し台センタの間で保持して行なわれて

center. The advantages of this method are quick set up and a positive drive.

いる．この方法の利点は，段取りが早くかつ駆動が確実なことである．

If a lathe has been used for taper **turning** with the tailstock offset, however, the tailstock may not have been realigned properly.

もし，旋盤が心押し台をずらしてテーパ**削り**に使われていたなら，心押し台は軸心が正しく出し直されていなかったかもしれない．

The tool and toolholder should not overhang too far for rough **turning**, but should be kept toward the toolpost as close as practical.

荒**旋削**には，バイトおよびツールホルダを出しすぎないようにし，刃物台のほうにできるだけ近く保持すること．

In longitudinal **turning**, the tool is moved parallel to the axis of rotation, so that cylindrical shapes are obtained. In transverse **turning** (also known as facing) the tool is moved at right angle to the axis.

縦（左右）方向**旋削**では，バイトは回転軸と平行に動かされるので，円筒形が得られる．横（前後）方向**旋削**（また，面削りともいう）では，バイトは（回転）軸に直角に動かされる．

The **turning** operation for the pinion gears in the productivity analysis was done on a Detroit tracer lathe.

生産性解析の一環として行なったピニオンギアの**旋削**作業は，Detroit ならい旋盤で実施した．

〈用　語　例〉

form turning　総形削り　taper turning　テーパ削り

under cut: to cutting a leg neck　**逃げ切り**

The **under cutting** tool is brought to the workpiece and the micrometer dial is zeroed. Cutting oil is applied to the work.

逃げ切りバイトを工作物まで動かし，マイクロメータのダイヤルをゼロにする．切削油を工作物にかける．

〈用　語　例〉

grinding under cut　アンダーカット研削

組立てる

assemble: to bring or come together; to fit or put together; to collect and put together the part of	**組立てる**

The system **assembles** two axles at a time.	このシステムは，一度に2つのアクスルを**組立てる**.
Lubricate parts with clean hydraulic fluid and **assemble** components with care to prevent damage to O-rings.	部品にきれいな作動油を塗り，Oリングを傷付けないよう気をつけて部品を**組立てる**.
The joints are easy to **assemble** and **disassemble**.	ジョイントは，**分解・組立**が容易.
Anything you **assemble can be assembled** automatically by a G. machine.	**組立てる**ものなら何でも，G機で自動**組立できる**.
Almost any product that is made in high volume **can be assembled** more efficiently with a G. machine.	G機を使えば，大量に作る製品ならまずどんなものでも，もっと効率良く**組立できる**.
Leave laminated disc assemblies **assembled** on forward short shaft.	前方の短かい軸に**組付けられた**多板ディスク組立品は，そのままにしておく.
Through operator assistance and automatic assembly stations, the joint **is assembled to** the axle.	ジョイントは，オペレータの補助と自動組立ステーションによってアクスルに**組立てられる**.
Mark blades and grips so they **may be reassembled** in the same combination.	同じ組合せで**ふたたび組立てられる**ように，羽根とグリップ〔固定金具〕に印をつける.

Machine parts **are** first **assembled** in units or subassemblies which **are** subsequently **assembled** in a final unit or assembly.

機械部品は，まず〔いくつかの〕ユニット，すなわち１次組立品に**組立てられ**，つぎにそれが最終ユニット，すなわち最終組立品に**組立てられる**．

Mark the intake pipes as they are removed from the engine so they **may be reassembled** in the same location from which they are removed.

吸入パイプには，それを取外したときと同じ位置に**ふたたび取付けられる**よう，それをエンジンから取外すときに印をつける．

Reassembling the gear box will be undertaken in the reverse order to dismantling.

歯車箱を**ふたたび組立てる**のは，分解するのと逆の順序ですることになる．

Fill groove in shaft with WANEN grease at **assembly**.

組立のとき，軸の溝にワネン・グリースを詰める．

Do not install shims and packing at this point of **assembly**.

組立のこの時点では，シムおよびパッキンは取付けない．

Lower strut and fork are a press fit, drilled on **assembly**.

下部支柱およびフォークは圧入で，**組立**のときに穴明けする．

Refer to figure 5 during **assembly** of wheel brakes.

ホイールブレーキを**組立てる**ときには，図５参照のこと．

Where **assembly** involves the handling of heavy components and where bearings are difficult to reach and/or see,……

重い部品も取扱う**組立**の場合，および軸受に手が届きにくかったり見えにくい場合には，……

This division makes custom-made **assembly** machine for the automotive industry,……, and any other application which requires the **assembly** of mechanical

この会社〔ディビジョン〕は，自動車産業，……その他多量生産で機械部品の**組立**を必要とする用途向け**組立**機械を受注製作

components in high volume production.

している．

The inside bearing inner ring with its **roller and cage assembly** should then be mounted on the axle.

つぎに，**保持器付コロ**の付いている内側軸受の内輪を，アクスルに取付けること．

〈用　語　例〉

assembler　アセンブラ，翻訳ルーチン，記号変換ルーチン

assembler language　アセンブラ言語

assembling clamp　クランプ締付機

ball and cage assembly　保持器付玉

general assembly drawing　総組立図

roller and cage assembly　保持器付コロ

build: to construct by putting parts or material together　（組立てて）つくる

The axle **was** then **rebuilt**, this time retaining the broken pinion bearings and gears, replacing only the carrier bearings.

つぎに，車軸を**ふたたび組立てたが**，このときは破損ピニオン軸受と歯車はそのままで，キャリア軸受だけ交換した．

The two test bearings **were built up** and assembled into a Skylab life test fixture using Belleville spring for axial load per Fig. 1 .

2つの試験軸受を**組上げて**，図1のように軸方向負荷に皿バネを使い，スカイラブ寿命試験フィクスチャ（取付け具）に組込んだ．

At station 15 the last operation in the output assembly **build up** is performed automatically.

ステーション15で，出力アッシ組立の最後の作業が，自動で**行なわれる．**

〈用　語　例〉

built up broach　組立式ブローチ

combine: to join or be joined into a group or set or mixture　組合せる，一緒にする

The key slotting tools **combine** tapered

このキー溝立削り工具は，勾

serrated HSS blades **with** a heat treated alloy steel holder.

配のある鋸歯付き高速度鋼の刃と熱処理した合金鋼ホルダを**組合せた**ものである．

The CNC machining center **combined with** the Rotary Indexer is a four-axis unit. It performs precision indexing rotary milling and helical milling.

回転割出し装置と**組合せた**このCNCマシニングセンタは，4軸ユニットである．これで，精密割出し，回転フライス削りおよびヘリカルフライス削りをすることができる．

On 2-axis numerically controlled machines the **combination of** the 4-speed AC motor and the 2-speed electric clutch provides 8 speeds per spindle.

2軸数値制御機械では，4速交流モータと2速電気式クラッチの**組合せ**によって，主軸は8速になる．

A **combination of** a variable-speed motor **plus** a mechanical form of gear change was found to be the best solution.

変速モータと機械的歯車切換え方式の**組合せ**が，最良の解決策であることがわかった．

connect : to join or be joined	**つなぐ**

Marine-turbines **are direct-connected** <u>to</u> the propeller shaft.

マリン・タービンは，プロペラ軸に**直結**である．

construct : to make by placing parts together ; to fix together from	**(組立てて) つくる**

It was therefore, necessary to **construct** a new test apparatus for use in this work.

それで，この研究に使う新しい試験装置を**つくる**ことが必要であった．

A spring-loaded 4-ball wear tester **was constructed**, that used 6.35mm diameter balls.

バネ負荷の4球式摩耗試験機を**つくった**．それには6.35mmの玉を使った．

けがく

layout: an arrangement of plan etc.	けがく，段取りする

Layout the 5 inch length from the end of the piece to the angle vertex.

加工物の端から角の頂点までの長さ5インチを**けがく**．

The hermaphrodite caliper can also be used to **layout** the center of round stock.

片パスは，丸材の中心を**けがく**のにも利用できる．

Angle may **be layedout** by placing the workpiece on the sine bar.

斜の線は，加工物をサインバーに乗せることによって，**けがく**ことができる．

Accomplish **layout** using the vernier height gauge.

バーニア付ハイトゲージ（高さゲージ）を使って，**けがく**．

Layout for stock cut off may involve a simple chalk, pencil, or scribe mark on the material.

材料を切断する場合の**けがき**には，簡単なチョーク，鉛筆またはけがき針による印などがある．

The sine bar can be used in angular **layout**.

斜線の**けがき**に，サインバーが使える．

Only enough pressure should be applied with the scriber to remove the **layout** dye and not actually remove material from the workpiece.

けがき針で**けがき**用塗料はとれるが，実際に工作物から材料を削らない程度の力を加えること．

mark: a line or object serving to indicate position	けがく

When scribing against a rule, hold the rule firmly. Tilt the scriber so that the tip

定規に当ててけがくときには，直尺をしっかり押さえる．けが

marks as close to the rule as possible. This will insure accuracy.

き針を，先が直尺にできるだけ近く**けがく**ように傾ける．こうすれば正確にけがける．

The hooked leg is placed against the round stock and an arc **is marked** on the end of the piece.

曲がった脚を丸材に当てて，その端面に円弧を**けがく**．

scribe : a person who (before the invention of printing) made copy of writings	**けがく，けがき針**

Carbide scribers are subjected to chipping and must be treated gently. They do, however, retain their sharpness and **scribe** very clean narrow lines.

超硬のけがき針は欠けることがあるので，ていねいに取扱うこと．そうすることで，シャープさを失なわないで，きわめて明瞭な細い線が**けがける**．

Set the height gage to 0.750 inch and **scribe** the height equivalent of the frame end thickness.

高さゲージを 0.750 インチにセットし，フレーム端面の厚さに等しい高さを**けがく**．

Perpendicular lines may **be scribed** on the workpiece by the following procedure.

つぎの要領で，加工物に，垂直な線を**けがく**ことができる．

The workpiece should **be scribed** for a short distance at this point.

加工物のこの点に，１本の短かい線を**けがく**．

How **are** perpendicular lines **scribed** with a height gage?

高さゲージで(を使って)，垂直な線を**けがく**方法は？

A surface gage may be used as a height transfer tool. The **scribe** is set to a rule dimension and then transfer to the workpiece.

トースカンは，高さを移す道具としても使える．**けがき針**を物指しの寸法にセットしてから，加工物に移す．

Trammel points are used for **scribing** circles and arcs when the distance involved exceeds the capacity of the divider.

ビームコンパスの脚は，デバイダでは届かない寸法のとき，円および円弧を**けがく**ために使う．

Scribe the centerline of round stock with the hermaphrodite caliper (scribing caliper).

片パスで，丸材の中心線を**けがく**．

研削する，とぐ

cut: to shave with tool　　**削る**

Friable and semifriable types, which **cut** faster and cooler with reduced grinding pressure, are generally preferred.

（砥材としては，）研削圧が小さく，かつ，研削が早くて温度が低く，しかも，破砕性が高いもの，あるいは中程度のものが一般に好まれる．

grind: to sharpen or smooth by friction　　**研削する（とぐ）**

grind sharply.

鋭く**とぐ**．

More skill is needed to off-hand **grind** a chip breaker on a high speed tool than is required to **grind** the basic tool angle.

高速バイトに手でチップブレーカを**研削する**には，基本的なバイト角度を**研削する**のに必要な技倆より以上の技倆が必要である．

For more precise requirements, it is recommended that parts **be** polish-**ground** (Contact Speed Fam Technical Services Dept)

より高精度を必要とする場合には，部品をポリッシ（磨き）**研削する**ことをおすすめします（Speed Fam 技術サービス部にお問い合わせ下さい）．

In still another example, in-process

さらにもう一つの例で，イン

gaging plays a big role in **grinding** <u>of</u> stainless steel balls for ball valves. These balls are **ground** <u>on</u> their <u>OD's</u> with a Heald universal internal <u>grinder</u> equipped with special tooling.	プロセスのゲージング（寸法測定）が球弁用のステンレス鋼球<u>の</u>**研削**で大きな役割を演じている．この球は，特殊なツーリングを備えたヒールド万能内面<u>研削盤で</u>，その<u>外径が</u>**研削される．**
Any batch of gears —spur, helical, external or internal— **ground** <u>on</u> Model G or 18inch gear grinder……	G形または18インチ歯車研削盤で**研削した**歯車——平，ハスバ，内，外——はどんな数のロットでも……
The running surfaces were **ground** to a surface roughness of 4 ~ 8 μinch CLR.	走行する表面を，表面粗さ4 ~ 8 μinch CLRに**研削した．**
Work material consisted of 25.4×152.4×304.8mm plates. Plates **were** <u>surface</u> **ground** to remove any oxide layers formed during cut off operation.	加工材は25.4×152.4×304.8 mmの板である．板は切断作業中にできた酸化層を除去するために<u>平面</u>**研削した．**
Grinding is done <u>by</u> line contact with the inner edge of the cup wheel, with the face of the wheel being dressed periodically by a single-point diamond.	定期的に砥石使用面を単石ダイヤモンドで目立して，カップ砥石の内側の縁と線接触<u>で</u>**研削する．**
Figure 1. shows the **grinding** <u>of</u> an 8″ diameter, flanged aircraft bearing race <u>on</u> a No.2 MICROCEN-TRIC shoe-type <u>grinding machine</u>, a product of Cincinnati Milacron.	図1は，シンシナチ・ミラクロン製MICROCEN-TRICシュー・タイプ<u>研削盤による</u>直径8インチフランジ付き航空機用軸受軌道輪<u>の</u>**研削**を示す．
The condition of the materials, the type of operation performed, variations in system parameters, and requirements with respect to tolerances and finish all affect	材料の状態，行なう作業の種類，システムのパラメータの違い，要求される公差および仕上げ程度などのすべてが**被研削性**

grindability.. に影響する．

〈用　語　例〉

centerless grinder　心無し研削盤
cutter grinder　工具研削盤
grindability　被研削性
grinding　研削
internal grinder　内面研削盤
multiple rib grinding wheel　多ネジ山砥石
plain grinder　円筒研削盤
surface grinding　平面研削
tool grinder　工具研削盤
unground bearing　非研削軸受

〈関連用語〉

abrasive　研削する，とぐ
abrasiveness　研磨性能（研磨力）
abrasive powder　微粉（砥粒の）
abrasive processing　砥粒加工
coated abrasive machining　研磨布紙加工

remove: to cut or shape　　**研削する**

A.T. Centerless **removed** 0.150 stock from a ¾″dia. cast iron shaft in only nine seconds, holding roundness within 0.0002.

A.T 心無し研削盤は，直径¾インチの鋳鉄の軸からわずか9秒で取代0.150（真円度0.0002以内で）を研削をする．

sharpen: to make or become sharp.
sharp (having a fine edge or point that is capable of cutting or piecing not blunt)　　**とがらす，とぐ**

Sharpen the knife <u>on</u> <u>a</u> <u>whetstone</u>.

<u>砥石で</u>ナイフをとぐ．

〈用　語　例〉

Sharp V-thread　シャープVネジ
sharpness　尖鋭度〔砥粒の〕

合成する，一緒にする

combine: to join or be joined into a group or set or mixture　　**合わせてつくる（結合，混合）**

Combine several simple vibrations with different frequencies.

周波数の異なるいくつかの単純振動を**合成する．**

This compound **is** then **combined with** cobalt, and the resulting mixture is compacted and sintered in a furnace at about 1,400℃.

つぎにこの化合物をコバルトと**混合**して，できた混合物を圧縮し，約1,400℃で炉の中で焼結する．

One-directional thrust bearings can support heavy thrust load in one direction, **combined with** a moderate radial load.

一方向スラスト軸受は，適度なラジアル荷重と**併わせて**，一方向の重スラスト荷重を支えることができる．

Many different shaped parts can be produced by upsetting or heading. When **combined with** other metal-forming operations such as extrusion, piecing, trimming and threading, possibility of the process are almost unlimited.

据込みや頭部成形によって，多数の異なった成形部品をつくることができる．押出し，打抜き，トリミング（バリ抜き）およびネジ成形のような，他の金属成形作業と**組合せた**場合には，できる作業はほとんど無限といってよい．

This system, used **in combination with** the anti-friction Teflon material on the bearing surfaces of the table, saddle and slide ways, greatly increases the life of the machine and minimizes down time.

テーブル，サドルおよび案内面の軸受表面に減摩擦性テフロン材を**併用した**このシステムによって，この機械の寿命を大幅に延ばし，動作不能時間をきわめて短かくすることができる．

Two boards **are combined** by three rivets.

2つの板は3本のリベットで**結合されている．**

〈用　語　例〉

combination　組合せ，結合
combination chuck　両用チャック，複同チャック
combination gauge　組合せゲージ
combination trap　組合せトラップ
gross combination weight　連結総重量

combined efficiency　総合効率
combined lathe　万能工作機械
combined stress　組合せ応力
combined carbon　化合炭素
combination tone　結合音
combined bearing　複合軸受
combined cutting　複合削り
combination grain size　混合粒度
combined pipe tap and drill　ドリル付タップ

synthesize: to make by synthesis
synthesis: the combination of separate parts or element to form a complex whole, artificial production of a substance that occurs naturally in plants or animal
組合せる，合成する

Synthesize an electronic circuit in computer.
コンピュータの電子回路を**合成する．**

Protons **are synthesized** from amino-acid.
プロトンは，アミノ酸から**合成する．**

こする，する（摩擦）

chafe: to warm by rubbing; to make or become sore from rubbing
こする，すりむく，擦過，摩擦

Subject tubes have been reported **chafing** against the tube guide.
標記チューブは，チューブ・ガイドを**こすって**いたと報告されている．

friction: the rubbing of one thing against another.; the resistance of one surface to another that moves over it
摩　擦

It reduces **friction** for cooler, quiet operation.
より低温・低騒音運転ができるよう**摩擦**を減らす．

The **friction** is considerably reduced by
摩擦は，切削液をかけること

the application of the fluid.

によって，かなり減る．

The presence of the welded material further increases the **friction**,……

溶着物があると**摩擦**がさらに増え，……

In any mechanical device, it is impossible to eliminate **friction**. Nether the less, in many respects, the working efficiency of the machine is governed by the degree to which the **friction** of its parts is lessened.

どんな機械・装置でも，**摩擦**を無くすることはできない．しかし，多くの面で機械の作業効率は，その部品の**摩擦**の減少度合いに左右される．

Sliding **friction** is caused by inequalities in the surfaces and the extent of **frictional** resistance is determined by the force of friction, or the force required to maintain a uniform velocity of one body with reference to the other.

滑り**摩擦**は表面の凹凸が原因で，**摩擦**抵抗の大きさは摩擦力，すなわち１つの物体が他の物体に対して等速を維持するのに要する力によって決まる．

……creat **friction**.

摩擦を生じる

Friction may be generated.

摩擦を生じる．

The slight deformation in the surface of one or both bodies which set up internal **friction**……

内部**摩擦**を発生させる片方あるいは両方の物体表面のわずかな変形は……

It determines the amount of sliding **friction** developed.

それで，発生する滑り**摩擦**の量が決まる．

Friction occurring on the cutting tool face……

バイトすくい面に起こる**摩擦**は……

In machining it has been accepted that the **friction** between the chip and tool is an important factor in determining the

切削で，切屑とバイト間の**摩擦**が，発生する切屑のタイプ，つまり，表面仕上げの質および

type of chip produced and, consequently, the quality of the surface finish and the efficiency of metal removal.

切削効率を決める1つの重要なファクタであるということはもう常識である.

Internal friction in fluids, as in the flow of fluid in a pipe where central portion is moving faster than the portion in contact with the wall, is called the force of viscosity.

パイプ中を流れる液体の中央部分が,壁に接触している部分よりも早く動いているような液体中の**内部摩擦**を,粘性力という.

The **friction** on the tool face is a most important factor in metal cutting.

バイトすくい面の**摩擦**は,金属切削で最も重要なファクタの1つである.

friction at the chip-tool interface is……

切屑・バイト界面の(における)**摩擦**は……

The static and break-away **friction** of PTFE which has been at rest is higher for the first movement than for subsequent movements.

静止状態にあったPTFEの静止およびブレーキ・アウェイ(急に動いたときの)の**摩擦**は,最初の運動時のほうがその後の運動よりも高い.

A high coefficient of **friction** results in a decrease in the shear angle and is the major factor leading to the formation of built up tool edge.

高い**摩擦**係数は剪断角を減少させ,構成刃先を形成するようにする重要なファクタである.

A considerable reduction in the mean **coefficient of friction** on the tool face can be achieved, under certain conditions, by the application of the correct lubricant.

適正な潤滑剤を使うことによって,特定の条件下で,バイトすくい面の平均**摩擦**係数をかなり低減することができる.

〈用 語 例〉

friction grip　摩擦接合用ボルト
friction sawing machine　高速切断機
friction screw press　(円板駆動形)摩擦プレス
friction welding　摩擦圧接

frictioning　フリクショニング，フリクション操作

rub: to press something against (a surface) and slide it to and fro.; to polish or clean by rubbing	**こする，こすりつける，摩擦する**

When the thrust force per unit back engagement (width of cut) is very low, only **rubbing** occurs, where the grains **rub** on the work surface with essentially no material removal.	切込み（研削幅）当りのスラストがきわめて小さいときは，ただ**こすっている**だけで本質的な研削はまったくなく，砥粒は工作物の表面を**こすっている**だけである．
Generally when different substances **are rubbed together**, electricity is generated.	異種の物質を**こすり合わす**と通常，電気が起こる．
Rub a glass bar with a silk cloth.	絹布でガラス棒を**こする**．
We know that lubrication is necessary to all points at which one surface **rubs** against another.	表面が他の表面と**摩擦する**点にはすべて潤滑が必要である，ということは皆知っている．
Rubbing occurs in bearings which support rotating shaft, in gears which have meshing teeth, and between pistons and cylinders in which they operate.	回転軸を支える軸受，歯が噛合っている歯車，シリンダとその中を動くピストンとの間には**摩擦**が発生する．
Before inserting a drill into a socket, **rub off** the shank to certain that it is smooth and free from grit.	ドリルをソケットに挿入する前に，滑らかでゴミがないことを確かめるためにシャンクを**ぬぐう**．
Rub the material against the wall with a stirring round bar.	攪拌丸棒で材料を壁に**こすりつける**．

〈用　語　例〉

colour fastness to rubbing　摩擦堅牢度　　rubbing compound　摩擦合成

固定する，止める（取付ける）

bolt：to fasten with a jig.	ボルトで固定する

…. Now **bolt** the rod and big end cap on to the pin. Rotate the crankshaft and take the bearing apart and examine the markings. The metal should be scraped away until the marking is evenly distributed all over the bearing.

ここで，コンロッドおよび大端キャップを（クランク）ピンに**ボルトで取付ける**．クランク軸を回転し，軸受を取外して，アタリを調べる．アタリが軸受全体に均一に分布するまで，メタルをキサゲでとる．

It may **be bolted** directly to the nose of the spindle.

それは，主軸ノーズに直接**ボルトで取付け**できる．

The work should **be** well clamped or **bolted** to the worktable.

工作物は，ワークテーブルに十分にクランプするか，**ボルトで固定する**こと．

…by **bolting** of the workpiece onto the worktable using the T slots provided.

既設のT溝を使い，工作物をワークテーブルの上にボルトで固定することによって，……

A means of **bolting** the workpiece to the faceplate…

工作物を面板に**ボルト止めする**方法は，……

There will be no slippage between **bolted** parts.

ボルトで固定した部品の間に，滑りはまったくない．

clamp：a device for holding things tightly often with a screw；to grip with a clamp；to fit firmly	締付ける，しっかり取付ける，クランプで固定する

……**is clamped** on the flange by the lock nut.

……を止めナットでフランジに**固定する**．

The center race should **be clamped** firmly against a true shoulder on the shaft by means of a distance sleeve and end nuts.

ディスタンス・スリーブおよび軸端ナットによって，中央輪を軸と一体のショルダーに当ててしっかり**固定する**．

Parting tools in common use include those of high-speed steel (HSS), of high cobalt alloy, with a brazed carbide tip, with either screw or pin-**clamped** carbide inserts, and more recently, with self-gripped carbide inserts.

普通使われる突切りバイトには，高速度鋼製，高コバルト合金製，ろう付け超硬チップ，ネジかピン**留め**の超硬インサート，そしてごく最近のものでセルフ・グリップ形超硬インサートなどがある．

When the half-nuts **are clamped** on the thread of the lead screw, the carriage will move a given distance for each revolution of the spindle. This distance is the lead of the thread.

ハーフ（半割り）ナットを親ネジのネジに**クランプする**と，往復台は主軸の各１回転に対してある決めた距離を動く．この距離がネジのピッチである．

Where the insert can **be clamped** onto a tool holder, ……

インサートをツールホルダに**固定**できる場合は，……

Parts can **be** held in fixtures, chucks, collets, vises or merely **clamped** directly to the table.

部品は，取付具，チャック，コレット，バイスで保持することも，単にテーブルに直接**固定することも**できる．

When **clamping** the work in place, only wrenches which properly fit the nut and boltheads should be used.

工作物を所定の位置に**固定する**場合には，ナットおよびボルト頭にぴったり合うスパナ以外は使わないこと．

Once the job is complete, the units can **be** simply **unclamped** and moved to another setup.

仕事が終われば，ユニットは簡単に**取外し**て，他の段取りにもってゆくことができる．

Positive locking fingers grip the tool adaptor (shown in locked and unlocked position). Tool is held in position by a series of Belleville springs and **unclamping is** accomplished hydraulically.

ポジティブ・ロッキング（確実固定）フィンガがツール・アダプタをつかむ（ロックおよびアンロックの状態を示す）．ツールを所定の位置に保持するのは一連の皿バネで，**アンクランプ**は油圧で行なわれる．

Reclamp with the lever.

レバーで**ふたたびクランプする**．

〈用　語　例〉

arm clamping mechanism　アーム締付機構
automatic clamping device　自動締付装置
band saw clamp　帯鋸クランプ台
casing clamp　箱締付機
clamp coupling　クランプ継手
clamped tool　クランプ工具
clamping bolt　締付ボルト
clamp screw　締付ネジ
hydraulic clamping mechanism　油圧固定機構
needle clamp body (needle stopper)　針止め
needle clamp complete　針止め組
needle clamp thumbscrew　針止めちょうネジ
rotating clamp　回転クランプ
strain clamp; anchor clamp　引留クランプ

fasten; to fix firmly, to tie or joint together　**固定する，取付ける，縛る**

Fasten the rope to the post.

ロープを柱に**縛りつける**．

Workpieces must **be fastened** securely for the machining operation.

切削作業では，工作物をしっかりと**固定**しなければならない．

The two most commonly used workholding methods on the vertical milling machine are the **fastening** of workpieces to the machine table and the holding of workpieces in a machine vise. When a workpiece **is fastened** to the machine table, it must be aligned with the axis of the table.

立フライス盤で最も普通に使われる工作物の２つの保持法は，機械のテーブルに工作物を**固定すること**と，機械バイスで工作物をつかむことである．工作物を機械テーブルに**固定する**ときには，工作物はテーブルの軸と心が，合っていなければならない．

…**fasten** or **unfasten**…to or from the instrument.　…で……を装置に**固定**したり**解き戻し**たりする．

〈用　語　例〉

fastening plate　止め金具

---

fix：to fasten firmly, to hold steadily, to set or place definitely.；to establish, to settle　**固定する，しっかり取付ける**

A bar with one end **fixed** and the other end free….　一端**固定**，一端自由のバーは…

The coil springs **are fixed together** with small metal clips.　コイルバネを小さい金属クリップで**止め合わせる**．

Fix…by tightening the clamping screws.　クランプネジを締付けて，…を**固定する**．

Fix…on a stable and heavy stand (base).　安定した重いスタンドに，…を**固定する**．

The mirror **is fixed** to the wall.　鏡は壁に**取付けられている**．

Fix a flask in a tilted position.　フラスコを傾けて**取付ける**．

…**is** firmly **fixed** at right angle to the frame.　…をフレームに直角にしっかりと**取付ける**．

Quenching **fixes** the structural change in the metal.　焼入れによって，金属の組織変化は**固定**される．

〈用　語　例〉

fixed angular table　固定角テーブル

fixed axle type　フレーム固定式

fixed beam　固定梁

fixed displacement pump　定容量形ポンプ

fixed end 固定端
fixed end cylinder 固定端シリンダ
fixed grain 固定砥粒
fixed load 固定荷重
fixed memory 固定記憶装置
fixed storage (real-only storage) 固定記憶装置
fixed word length 固定語長
fixing 固着
fixture 取付具

key: fixed with key　**キーで固定する**

The crankshaft and camshaft timing wheels **are keyed** on to their shafts.

クランク軸およびカム軸のタイミング歯車は，軸に**キーで固定する**.

lock: to fasten or to be able to be fastened with a lock ; to bring or come into a rigidly fixed position　**ロックする，止める，固定する**

A quarter turn of the lockcup is all that is needed to **lock** the insert.

インサートの**固定**で必要なことは，ロックカップを1/4回転させることだけ.

The head and column clamp should always **be locked** when drilling is being done on a radial drill press.

ラジアルボール盤で穴明けしているときは，ヘッドおよびコラムのクランプは常に**ロック**しておく.

…, by **unlocking** the tool-post with the lever provided.

備え付けてあるレバーで，刃物台を**緩める**ことによって，……

Method of adjustment at this point varies slightly with different cars. There may be a cap which screws down as shown in Fig. 360, and which **is locked** in position by a small grub screw, or there may be the well-known threaded ball-socket as shown in Fig. 361. This **is locked** with a split pin which engages with the screwdriver slot in

ここの調整方法は車によって若干違う. 図360のようなねじ込みキャップがあるが，これは小さいグラブネジで所定位置に**固定する**. あるいはまた，図361のような，よく知られているネジ・ボールソケットもある. これはソケットの頭のネジ廻し用

the head of the socket.

すり割りにはまる割りピンで**固定する.**

Vee-Flange adaptors **are** solidly **locked** into the spindle taper by the fingers shown in the accompanying illustration.

Vフランジ・アダプタを，図示のように指でスピンドルのテーパに固く**固定する.**

The turret **is locked** solidly to the slide gear box by means of a Vee-type peripheral clamp ring that clamps a full 360 degrees around the back of the turret and the front of the slide gear box (Refer to cross sectional illustration.). The clamp ring **locks** the turret to the slide; there is no reliance on the shot pin to hold the turret into position.

タレットは，タレット背面とスライド・ギアボックス前面の周りを360°にフルにクランプするV形円周固定輪によって，スライド・ギアボックスに堅固に**固定される**(断面図参照)．この固定輪がタレットをスライドに**固定する**；タレットを所定位置に保持するために，ショット・ピンにはまったく頼らない.

At this position, you **lock** the indicator to the spindle of the pedestal.

この姿勢で，インジケータをペデスタルのスピンドルに**固定する.**

The arm and head can **be** raised or lowered on the column and then **locked** in place.

アームおよびヘッドは，コラム上を上げ下げして，所定位置に**固定**できる.

The ring can **be locked** in a specific position.

リングは，指定位置に**固定**できる.

The table consists of a base and X and Y axis tables which move into position by air cylinders, and which also **lock** the table against X and Y stops when the table is positioned for a machining operation.

このテーブルは，ベースおよびX軸テーブルとY軸テーブルで構成されていて，エアシリンダで所定位置に動き，切削作業を行なうために位置決めされると，テーブルはエアシリンダで

Xストッパおよびストッパに当てて**固定される**.

The cassette **is locked** into position against hardened locator pin for constant alignment.

カセットは，一定したアライメントが保てるように，焼入れ硬化した位置決めピンに当てて，所定位置に**固定される**.

The operator slowly moves the welding head until it's in position, then signals the computer. The location becomes **locked** in the memory.

溶接ヘッドを所定位置となるまでゆっくり動かし，それからコンピュータに信号を送る．その位置が記憶装置に**ロックされる**ようになる.

〈用　語　例〉

bracket lock spring　ハンガー止めナット
lock nut　止めナット（check nut）
lock washer　止め座金
locking bolt　締付ボルト
self-locking nut　戻り止め（カシメナット，ファイバ入り）ナット
toothed lock washer　歯付座金

screw：to fasten or tighten with a jig　　**ネジで固定する**

Sight glasses **are screwed** directly **into** the bottom of the bearing housing and no drain line is used.

サイトグラスを，軸受ハウジングの底に直接**ねじ込んで取付け**，ドレーンラインは使ってない.

This compression tester is a useful device which can **be screwed into** the cylinder in place of a sparking plug.

この圧力テスタは有用な装置で，点火プラグの代わりにシリンダに**ねじ込んで取り付け**できる.

〈用　語　例〉

screw auger　ボートギリ
screw conveyer　ネジ式コンベヤ
screw cutting　ネジ切り
screw cutting machine　ネジ切り盤
screw driven planer　ネジ式平削り盤
screw head　ネジ頭
screw jack　ネジジャッキ
screw motion　ネジ運動
screw plug　ねじ込みプラグ
screw pump　ネジポンプ
screw shaft　プロペラ軸

screw driven shaper　ネジ式形削り盤
screw gear　ネジ歯車
screw stock　ネジ用棒材
screw wrench　モンキースパナ
screw tap　ネジタップ

seat：to cause to sit, to put (machinery) on its support.　**（台座などに）止め付ける，固定する，据える**

If the live center is too small for the lathe spindle taper, use a tapered bushing that fits the lathe. **Seat** the bushing firmly in the taper and install the center. Make sure the bushing **is** firmly **seated** in the taper.

もしライブ・センタが旋盤主軸のテーパに小さ過ぎる場合，その旋盤にぴったり合うテーパのブシュを使う．このブシュをテーパにしっかりと**止め付けて**，センタを取付ける．ブシュがしっかりとテーパに**止め付けられている**ことを確める．

The feed pressure when drilling is usually sufficient to keep the drill **seated** in the tailstock spindle,thus keeping the drill from turning.

通常，穴明けのときの送り圧により，ドリルを心押し台に十分**止め付け**ておくことで，ドリルは回わらなくなる．

On cam-locking holders, the insert is placed over a pin which, when tightened, **seats** the insert against the back surface of the pocket.

カムロック・ホルダでは，ピンの上にあり，ピンを締めると，インサートをポケットの背面に当てて**止め付けられる**．

Another point to note is that the bonnet shut down correctly： it should **seat** down on to the damper-tape clear of the edge of the radiator.

もう1つ注意すべき点は，ボンネットを正しく閉めること．すなわち，ボンネットがラジエータの縁に触れることなく，ダンパテープの上に**すわる**ことである．

〈用　語　例〉
minimum seating stress　最小ガスケット締付圧

secure：to fasten securely　**固定する，取付ける**

Camshaft removal—Before the camshaft can be withdrawn it may be necessary to remove one or more set screws **securing** the bearings.	カム軸の取り外し—カム軸を取り出せるようにするには，それに先立って軸受を**固定している**１つあるいはそれ以上の止めネジを取外すことが必要になる．
Temporarily **secure** with two bolts and washers.	２つのボルトおよび座金で，仮に**取付ける．**
Secure a gear wheel to its shaft, by means of a key and key-way.	キーとキーウェイで歯車を軸に**固定する．**
The magnetic chuck was subsequently used on other machines for parts difficult to **secure** by mechanical means.	機械的手段で**固定する**ことの困難な部品には，その後他の機械で磁気チャックを使用した．
Where a whole tube cannot be got into the position, a split tube with its two halves **secured** together temporarily by wire, or even half tube only,may be employed.	筒のままでは入れることのできないところでは，２分割したものをワイヤで仮に一緒に**縛った**割り筒か，または半身の筒だけで使うことができる．
A chuck **secured** on unkeyed shaft…	キーなしの軸に**しっかり取り付けた**チャックは……
An end mill **is secured** to the arbor.	エンドミルは，アーバーに**固定される．**
Hydraulics **secure** the tailstock ram in position when workpieces are locked between centers.	工作物をセンタ支持で固定する場合は，油圧装置で心押し台ラムを所定位置に**しっかり固定する．**
A dog **is secured** at the headstock end.	回し金を主軸台端部に**しっかり固定する．**

さがす（捜査，探査），調べる

research：careful study and investigation, especially in order to discover new facts or information, to do research into	調べる（調査）

The subject **has been** fully **researched**.

このテーマは，**調べ**尽くされている．

〈用　語　例〉

research furnace　研究炉　　research worker　研究員

search：to look or go over (a place etc.) in order to find something, to examine thoughtfully, to look through	調べる，さがす（探索）

The **search** for effective laboratory test methods for evaluating cutting fluid performance has been in progress for over thirty years.

切削液の性能評価に効果的な，研究室における試験方法の**調査**は，30年以上にわたって続けられている．

〈用　語　例〉

binary search　2分検索　　search light　探照灯

searcher　検査器

seek：to make a research of inquiry for, to try to find or obtain	さがす（探求）

The researchers **sought** the cause of defects in cold-formed 4027 steel preforms used for automotive transmission gears.

調査員は，自動車のミッション歯車に使われている冷間成形4027鋼粗成形品の欠陥の原因を**探求**した．

With this system all vital points in the

このシステムでは，重要な個

installation are continually monitored **seeking** any Off-Normal or Change of State point.

所すべてを，異常や状況変化を継続的にモニターして**探し出す．**

The machining characteristics are **sought** for maximum productivity on shapers.

形削り盤が最大の生産性を得られるよう，その切削特性を調べる．

支える，支持する，担う

bear : to carry, to support　　**担う，支える**

Linear movements **are borne** on hardened and ground ball tracks, and the carrier spindle rotates in angular contact bearings.

直線運動は，焼入れ，研削されたボールトラックで**支えられ，**キャリア・スピンドルはアンギュラー軸受の中で回転する．

〈用　語　例〉

bearing metal　軸受メタル
bearing pressure, ground pressure　支え圧，軸受圧力
bearing shim　軸受はさみ金
bearing spring　支えバネ，軸バネ
bearing stress　支え応力
bearing surface　支え面
rolling bearing　転がり軸受

carry : to support the weight of, to bear　　**担う，支える**

Combinations of ball and roller bearings mounted on precision-bored housings can be used to **carry** journal and thrust loads of work spindles.

精密に中ぐりしたハウジングに取付けた玉軸受とコロ軸受の組合せを，主軸のジャーナルおよびスラスト荷重を**支える**ために使うことができる．

The primary shaft **is** usually **carried in** a large double-ball or roller bearing.

主軸は普通，大形の複列玉軸受あるいはコロ軸受で**支持する．**

The spindle **is carried in** a head that is vertically adjustable on the column.

主軸は，コラム上を縦方向に調整できるヘッド（主軸台）に**支持されている．**

A floating table **carried** on linear roller bearings in a square slideway, is located between the two heads.

四角の滑り案内の中の直線コロ軸受で**支えられている**浮動テーブルが，2つのヘッドの間にある．

The gear casing **is carried by** three struts within the tunnel.

ギアケースは，トンネル内で3つの支柱で**支えられ**ている．

〈用　語　例〉

carriage　往復台

carrier　回し金，ドグ

carrier plate　親板（板バネ）

carry-over loss　持逃げ損失

prop: to support or as if with a prop, to keep from falling or failing

（棒で）支える

Support for the hoop, to prevent it springing while the tacks are being driven in, can be provided by **propping** a length of wood under it, the lower end of the **prop** resting on the boards under the seat squab, or between the seats in the case of bucket seats.

鋲を打込んでいる間，フープが弾けないように，適当な長さの木をその下に**つっかい棒にする**ことによって，フープを支えることができる．この**支柱**の下端はシート座布団の下，あるいは折りたたみ式1人用シートの場合は，シートの間の板の上に静置する．

Prop the windscreen open while the screws by which it is attached to the frame of the window are removed.

風防ガラスを取付けている窓枠のネジを取外す間，風防ガラスを**支柱で支え**て，開いておく．

〈用　語　例〉

prop post　打柱（うちばしら）

support: to keep from falling or sinking
: to hold in position
: to bear the weight of

支える，支持する

It consists of a horizontal bed **supporting** the headstock, the tailstock, and the carriage.

それは，主軸台，心押し台と往復台を**支える**水平のベッドで構成されている．

……**is supported at** the other by a center.

……は，他端がセンタで**支えられ**ている．

The spindle **is supported by** bearings.

主軸は，軸受で**支えられ**ている．

The arbor **is supported in** a bearing.

アーバーは，1つの軸受に**支えられ**ている．

This machine consists of a cutting **tool** mounted on a boring bar that is **supported on** bearings outside the cylinder.

この機械は，シリンダの外側の軸受に**支えられ**た中ぐり棒に取付けてある切削工具で構成されている．

It is difficult to **support** the workpiece **between** centers.

この工作物を，両センタで**支えること**はむずかしい．

An air cylinder raises and lowers the slide. A wheel support table **is** also **supported from** a vertical slide which, in turn, is attached to the hold-down slide.

エアシリンダでスライドを上げ下げする．またホイール支持テーブルは，立スライドによって**支えられ**，立スライドはさらに圧下スライドに取付けられている．

The test bearing **is supported** on each side by three large, grease-lubricated ball bearing (a total of six) mounted in fixed housings. The drive mechanism transmits motion through the support and test bearings.

試験軸受は，固定ハウジングに取付けた3つの大きなグリース潤滑玉軸受によって，両側が**支えられている**．駆動機構は，支持軸受および試験軸受を介して運動を伝達する．

〈用　語　例〉

supporting point　支点

support program　支援プログラム

rear axle support　後車軸支え
supporting electrolyte　支持電解質
support bar　支え棒
support spring　支えバネ
work arbor support stand　ワークアーバ支えスタンド
work support blade　支持刃

sustain : to support　**支える**

Thrust bearings **sustain** no radial load whatever.

スラスト軸受は，ラジアル荷重をまったく**支え**られない．

The load **is sustained by** the balls at their two points of contact with their inner and outer races respectively.

荷重は，内輪および外輪のそれぞれに2点で接触している玉で**支え**られる．

差し込む，取付ける

insert : to put (a thing) in or between or among　**差し込む，間に取付ける**

Before **inserting** the arbor into the spindle or adaptors, be sure both arbor and spindle holes are clean and free from nicks.

アーバーを主軸あるいはアダプタに**挿入する**前に，アーバーおよび主軸の穴がきれいで欠けがないかを確かめる．

Shank is designed to **be inserted** in a corresponding taper hole in the end of the spindle.

シャンクは，対応する主軸の端にあるテーパ穴に**差し込める**ように設計されている．

The tangue **is inserted** in a slot at the bottom of the hole in the spindle.

タングを，主軸穴の奥にある溝に**差し込む**．

The individual **inserts** can either be brazed onto or clamped onto a steel body or tool holder.

個々の**インサート**は鋼のボディあるいはツールホルダにろう付けするか，またはクランプすることができる．

〈用　語　例〉

inserted chaser die　植刃ダイス，入刃ダイス
inserted chaser tap　植刃タップ，入刃タップ
inserted drill　植刃キリ
inserted tooth cutter　植刃フライス
insertion　差込み，挿入物
wire inserted asbestos yarn　金属線入り石綿糸

load：to put a load in or on, to fill with goods or cargo etc.　**取付ける，装塡する，積む**

…. This system will index and **load** a tool in 4 seconds.

…．このシステムは，ツールを4秒でインデックスして**取付ける**（装着する）．

The machine has six stations. The cylinder rods **are** manually **loaded** at station one. They flow through the machine in a horizontal position on vertically adjustable, hardened rails. Upon reaching work stations, they **are** hydraulically **loaded** in adjustable fixtures and clamped in place.

この機械は6ステーションのものである．ステーション1でシリンダ・ロッドを手で**装着する**．ロッドは，上下方向可変の焼入れ硬化したレールの上を水平姿勢で通る．加工ステーションに着くと，ロッドは可調整取付具に，油圧で**装着され**，所定位置に固定される．

Up to 30 different tools can **be loaded** in the magazine.

このマガジンには，いろいろなツールを30個まで**装塡**できる．

Up to 8 hour supply or stock can **be loaded** in the bar feed magazine.

このバー送りマガジンには，材料を8時間分まで**装着**できる．

All station 10 operations are manual. The operator **loads** a matched set of output and transfer gears into part nests on the pallet. The output shaft is then taken from its part nest, fitted with an oil seal, and added to the transmissions assembly.

ステーション10の作業はすべて手作業である．作業者が，出力歯車と伝達歯車を合わせた1セットをパレットの部品巣箱に**乗せる**．つぎに，出力軸をその部品巣箱から取出し，オイルシ

ールをはめて，ミッション組立部品に取付ける．

The tool changing arm automatically changes tools up to 20lbs. in weight to 5″ in diameter and 17½″ in length. Longer tools can **be** manually **loaded** into the spindle.

ツール交換アームは重さ20 lbs.，直径5インチ，長さ17½インチまでのツールを自動交換できる．これよりも長いツールは，手でスピンドルに**装塡**する．

6 assemblies can **be loaded** on the machine.

この機械には，6個のアセンブリを**取付ける**ことができる．

The bar stock **is loaded** from the back of the headstock through a hollow spindle.

棒材は，中空主軸を通して工作主軸台の後部から**装塡する**．

The only manual operations are **loading** and **unloading** the conveyor, and spray painting.

手作業は，コンベヤへの**ローディング/アンローディング**と噴霧塗装だけである．

The system is designed for automatic **load／unload** by robot or mechanical part handlers. For medium-volume operations, a multi-axis programmable robot can **load** one chuck and **unload** the other.

このシステムは，ロボットまたは機械的パーツ・ハンドラ（部品取扱い具）による自動**ローディング/アンローディング**用に設計してある．中量生産のための，多軸プログラマブル・ロボットは，一方のチャックで**取付け**，もう一方で**取外し**をする．

The machine code can **be loaded** into the computer.

機械コードをコンピュータに**ロード**できる．

Programs can **be loaded** manually through the front panel keyboard or by paper or magnetic tape.

プログラムは，人が前面のパネル・キーボード，あるいは紙テープまたは磁気テープで，**ロード**できる．

〈用　語　例〉

automatic loader and unloader ローダ/アンローダ

loading　装塡

load point　ロードポイント

unload　取外す

unloader valve (pressure regulating valve)　アンローダ弁

unloading　積降し

plug in：to connect electrically by inserting a plug into a socket　**差し込む，取付ける**

The explosion-proof L. that is perfect for hazardous atmosphere applications and even **plugs** right **in** where you have been using the other H.

防爆形Lは，危険な雰囲気で使っても完璧．他のHを使っているところでも，ぴったり**差し込みで取付けられる．**

The system's computer **plugs into** whatever type of computer being used.

このシステムのコンピュータは，今，使っているどんなタイプのコンピュータにでもプラグで**接続**できる．

When the hose **is plugged** into any of the inlet,…

ホースがインレット（入口）のどれかに**差し込まれている**ときは…

As can be seen from Fig. 1, this system consists essentially of a measuring roll, electronics,…. All units are connected via **plug-in** contacts.

図1でわかるように，このシステムは主として測定ロール，エレクトロニクス…で構成され，ユニット（装置）はすべて**プラグ・イン**式接触器を介して連結される．

The L. **plug-in** switch line is for applications that need rugged, enclosed switches but not the performance characteristics of a traditional machinetool type limit switches.

L. **プラグ・イン**（差し込み）スイッチ系列は，頑丈な密封式スイッチを必要とする用途向けで，**動作特性はこれまでの機械**式リミットスイッチとは異なる．

仕上げる

burnishing : to polish by rubbing　　バニシ仕上げ

The steel ball **burnishes** the hole to the required diameter.

鋼球で(を通して)，穴を所要の直径に**バニシ仕上げする．**

〈用　語　例〉
burnishing action　　バニシ作用（バフ加工の）

〈関連用語〉
buffing　　バフ仕上げ，バフ研磨，バフ加工

dress : to finish or treat the surface of　　仕上げる

Dress splines with fine Indian Stone if necessary.

必要なら目の細かいインディアン・ストーン(油砥石)で，スプラインを**仕上げる．**

〈用　語　例〉
saw tooth side dresser　　ノコ刃側仕上機
dressing　　目立て

file : to shape or smooth with file　　ヤスリ仕上げ

To **file** a flat surface, change the direction of strokes frequently to produce a cross-hatch pattern.

平らな面を**ヤスリ仕上げする**には，綾目ができるようにストロークの方向を頻繁に変える．

When **filing** on a lathe, use a low speed, long strokes and **file** left handed.

旋盤で，**ヤスリ仕上げする**ときは，ゆっくり，長いストロークで，左手で**ヤスリをかける．**

finish : to put the final touches to　　仕上げる

Finish the metal surface neatly.

金属表面を，きれいに**仕上げる．**

Flat finish endmill reportedly **finishes** hardened steel and exotic metals to perfectly flat surface finish as good as 20 rms.

仕上げ用フラット・エンドミルで，焼入れ鋼および新種の金属を，20rms のような完全に平らな表面に**仕上げられる**という．

Then after hardening, it **was finished** to exact size on his gear-cutting grinding machine and bevel gear generator.

焼入れしてから，彼の歯切り研削盤および傘歯車歯切り盤で，正確な寸法に**仕上げた**．

Finish the last 1/4 inch by hand if approaching a shoulder.

肩に近づいたら，手で，最後の1/4インチを**仕上げる**．

This manufacturer **finished** the seatings by hand scraping.

このメーカーは，手でキサゲして，座を**仕上げた**．

The machine **finishes** six turbine blades from tip to root in one continuous operation.

この機械は，６個のタービンの羽根を先端から根元まで，１回の連続作業で，**仕上げる**．

The **finishing** of machined surface with a file and abrasive cloth should not be necessary if the tools are sharp and honed

バイトが鋭く，よく研いであり，また送り，切削速度および切込みが適正なら，切削加工面

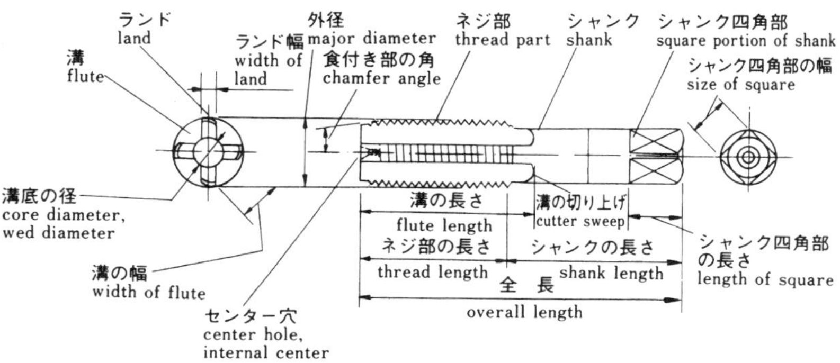

and if the feeds, speeds and depth of cut are correct.

をヤスリや研摩布で**仕上げる**必要はない.

For most purpose 0.002 inch is sufficient material to leave for **finishing**.

多くの場合, **仕上げ**に0.002インチ残せば十分である.

The groove tracks in a single-row ball bearing are **finish**-ground with grinding wheel.

単列玉軸受の溝軌道は, 砥石で, **仕上げ**研削する.

Surface **finish is** usually accomplished by grinding, lapping or honing.

表面**仕上げ**は, 通常, 研削, ラッピングあるいはホーニングでする.

To meet the demand for stronger, quieter gears for cars during the 1930's, the use of precision gear-grinding machines using formed wheel to **give** a precision **finish** to gear after hardening, increased.

1930年代, 自動車の歯車はより強く, より低騒音という要望に合うよう, 焼入れてから歯車を精密に**仕上げられる**総形砥石を使った精密歯車研削盤の使用が増えた.

…**generate** a 2 microinch **finish** on surface ground to about 7 microinches.

……によって, 約7 μインチに研削した面を, 2 μインチに**仕上げる**.

Use of cutting oils will help produce better **finishes** on finish facing cuts.

切削油の使用は, 仕上げ面削りで, より良い**仕上げ面を**作るのに役立つ.

In these figures, their surface **finish** behavior is illustrated by the surface **finish** by the rough form tool.

これらの図では, その表面**仕上げ**の様相を, 荒加工用総形バイトによって作られた表面**仕上げ**で示してある.

A is more machinable than B. This means that a better surface **finish** can **be achieved** with material A.

AはBよりも機械加工しやすい. このことは, 材料Aがより良い表面**仕上げが得られる**,

ということである.

Surface **finishes** of the order of 0.1μ have been obtained when carbide tips are ground.

超硬チップを研削した場合には，0.1μ のオーダ（桁）の表面**仕上げ**が得られている.

〈用　語　例〉

finish cutting　精密削り
finish forging　仕上打ち
finish marks　仕上記号
finished bolt　仕上ボルト
finished nut　仕上ナット
finishing　仕上げ
finishing punch　仕上パンチ
finishing roll　仕上ロール
finishing shop　仕上工場
finishing 〔hand〕 tap　仕上タップ
finishing tool　仕上バイト

honing：to sharpen on a hone (a fine-grained stone used for sharpening razors and tools)　**ホーニング仕上げ**

A slight flat should **be honed** on the end by hand with an oil stone.

オイル・ストーンを使って，手で端面に小さい平面を**ホーニング仕上げ**すること.

The bore **is honed** to a surface finish of 10.20μm.

穴は，10.20μm の表面仕上げに，**ホーニング加工**する.

Honing the edge of hand scraper blades requires a high degree patience and skill.

ハンド・スクレーパの切れ刃の**ホーニング仕上げ**は，高度の忍耐と技倆を必要とする.

Gear **honing** was originally developed as a method of inexpensively and surely removing nicks and burrs from hardened gear tooth.

歯車の**ホーニング**は，もともとは焼入れした歯車の歯からキズやバリを安く確実に除去する方法として，開発されたものである.

Another experiment involves the **honing**

一方の実験には焼入れしたポ

of blind-end bore in a hardened pump body.

ンプ本体の盲穴の，**ホーニング**もある．

Today there are two entirely different internal **honing** processes. One……, in the other the operator holds the workpiece, stroking it manually back and forth along honing tool.

現在，2つのまったく異なった内面**ホーニング**法がある．1つは……，もう1つの方法では，作業者が工作物を持って，ホーニング・ツールに沿って前後に手で往復させる．

〈用　語　例〉

free expansion honing　定圧ホーニング
honing head　ホーニング・ヘッド
honing stick　ホーニング砥石
vertical honing machine　立形ホーニング盤

lapping : to take up (liquid) by moving of the tongue, as a cat does

ラップ仕上げ，ラッピング

The spindle bar outside diameter **is lapped** for close fit to nitride-hardened steel bushing fitted in the sleeve.

スピンドルの外径を，スリーブにはめ込んである窒化焼入れ鋼製ブッシュにぴったりはまるように，**ラップ仕上げする**．

Figure 6 presents replication electron micrographs of the (110) plane of a head **lapped** by 14 μm Al_2O_3.

図6は，14μm Al_2O_3で**ラップ仕上げした**ヘッドの(110)面のレプリカ電子顕微鏡写真である．

〈用　語　例〉

lapping burn　ラップ焼け
flat lapping type　平面加工形
lapping force　ラッピング抵抗
lapping liquid　ラップ液
lapping machine　ラップ盤
lapping powder　ラップ剤

reaming : to finish of bore

リーマ仕上げ

Typically, a tool cuts the seat, a back-off mechanisim retracts the tool and a sqwirt-out reamer is advanced **to ream** the

バイトが座を切削，引き戻し機構がバイトを引込める，油穴付リーマが弁案内孔を**リーマ仕

valve-guide hole.

上げするように出る，というのが普通である．

Ream inside dia of bushings to 0.312-0.313 (Holes to be square with respective housings within 0.001 inch).

ブッシュの内径を，0.312～0.313に**リーマ仕上げする**(穴はそのハウジングに対し直角度0.001インチ以内のこと)．

When drilled holes have excessive eccentricity, they **are** bored 0.010 to 0.015 undersize and machine **reamed**.

穴明け加工した穴が過度に偏心しているときには，それを0.010～0.015だけ小さい寸法に中ぐりして，**機械リーマ仕上げする**．

Chucking **reamers** cut on the chamfer at the end of the flutes. This chamfer is usually at a 45 degree angle.

チャッキング**リーマ**は，溝の端の食付き部で削る．この食付き部は，通常45°である．

〈用　語　例〉

reamer bolt　リーマボルト
reaming　リーマ通し
reamer section tap　リーマ付タップ

scraping：to make (a thing) clean or smooth or level by passing the hard edge of something across it, to remove doing this

キサゲ仕上げ

Produce a clean or smooth finished surface by **scraping** away superfluous material.

加工物の余計な材料を**キサゲでとって**，きれいで滑らかな仕上げ面にする．

Carefully **scrape** the impurity from the surface.

よく注意して，表面から不純物を**キサゲでとる**．

shave : to cut or scrape thin slices from the surface of (wood etc.)　　シェービング仕上げ

We cold-formed it. Then we drilled the large center hole, **shaved** the hex.top, deburred and plated. That's eight secondary operations per parts.

それを冷間鍛造し，つぎに大きなセンター穴をあけ，六角頭を**シェービング**し，バリ取りしてメッキした．これは，部品毎に8回行なう，2次加工である．

The process of **shaving** as a means of giving a fine finish to unhardened gears was begun by the Machine Tool Company of Illinois.

焼入れしていない歯車を，上質の仕上げ面にする手段としての**シェービング**仕上げ法は，イリノイ州の工作機械会社が始めたものである．

These early gear-**shaving** <u>machines</u> use a racktype cutter.

これら初期の歯車**シェービング**<u>盤</u>は，ラック形カッタを使っていた．

〈用　語　例〉

gear shaving　歯車シェービング仕上げ
gear shaving machine　歯車シェービング盤
internal shaving attachment　内歯車シェービング装置
shaving cutter head　シェービング・カッターヘッド
shaving press　シェービング・プレス
shavings　削り屑

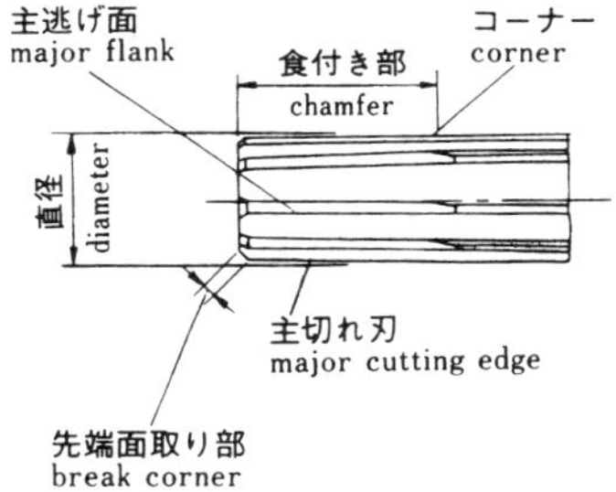

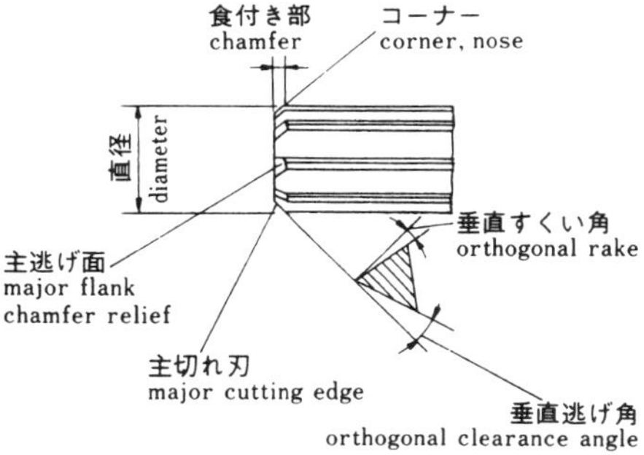

skive: to cut off, as leather, rubber, etc, in thin layer or piece : to shave or pare as hide or leather	**スカイビング仕上げ**

Quite often large moulded billets **are skived** or shaved into thin tape for use as liner or sliding flat surface.

大きな成形ビレットを，ライナや平らな滑り面として使用するために，薄いテープに**スカイビングする**ことがよくある．

してみる，やってみる，試みる

attempt: to make an effort to accomplish	**やってみる，つと（務）める**

The operator should **not attempt** to clean or oil the machine or make any adjustment to the work.

オペレータは，機械を掃除や給油し**ない**こと．また，工作物に対し，どのような調整もし**ない**こと．

It is well known that perhaps the earliest **attempt** to measure machining performance **was made** by F.W. Taylor near the beginning of this century to improve tool steel.

工具鋼を改良するため，今世紀の初め頃，切削性能を測定する最初の**試み**が，F.W.テーラーによって**なされた**ことは周知の通りである．

Where a fit proves to be tighter than specified, **make no attempt** to reduce the shaft seating by scraping it or by rubbing it with emery cloth or other similar material.

はめあいが指定よりもきついことが明らかになった場合，軸座にキサゲをかけたり，研摩布やこれと似た他のものでこすって，細く**しようとしない**こと．

try: to attempt, to make an effort to do something to test	**してみる，試みる**

Try to design the component so that machining is not needed on the unexposed surfaces of the workpiece when the component is gripped in the work-holding device.

工作物保持具で部品を把持したときに，工作物の露出していない面は切削する必要がないよう，部品を設計**してみる**.

Should the piston be pushed upward in this way **for purposes of trial** it will be desirable to remove at least one of the piston rings, since this will have been pushed beyond the top of the bore. If the ring is not removed **do not try** to pull the connecting rod down again without first compressing the top rings, for that would probably result in the rings being broken.

試しに，ピストンをこのように上方に押してみると，ピストン・リングが穴の上面を越えているから少なくともその1つを取外したほうがよい．もし，リングを取外していないなら，トップ・リングを圧縮しないで，ふたたびコンロッドを引き下げようと**しない**こと．なぜなら，そうすると，多分リングを破損することになるからである.

Initially, several combinations of ball separator materials **were tried** with the RF sputtered MoS_2 coatings.

最初玉軸受保持器材料の数種の組合せをRFスパッタのMoS_2コーティング後**試験**した.

To **try** ACIO <u>on</u> your job, or for more information, call or write S. Inc.

仕事<u>に</u>**試してみる**ため，またもっと情報を得るために，弊社に電話か手紙を下さい.

If it is not the case, the correct timing can be found by **trial and error**.

もし，そうでない場合には，正しいタイミングは，**試行錯誤**で見い出すことができる.

処理する

attend to : to apply one's mind
: to deal with (matter)

処理する，留意する

……. It is only a 15 minutes job, however, to **attend to** them all. All greasing should be carried out systematically.

……．しかし，全部**処理する**のは，わずか15分の仕事である．すべてのグリーシング（グリースの給油）は系統だてて行なうこと．

It is wise to disconnect one battery lug before **attending to** the cut-out.

カット・アウトする前に，バッテリの端子の１つの接続をはずしたほうがよい．

When all pistons have **been attended to** in this way proceed to scrape away the carbon deposits adhering to those portions of the cylinder-block which form part of the combustion chamber.

このようにピストンを**処置し**終わったら，燃焼室部分を形づくっているシリンダ・ブロック部分に固着しているカーボンの付着物を掻き取るように，作業を進める．

……, and this must **be attended** before they are reassembled.

……，それらを再組立てする前に，これを**処置し**なければならない．

cope with：to deal successfully with, manage successfully

うまく処理する，こなす

This self-lubricating material, a glass-impregnated graphite, should be to **cope with** the load and temperature transients one would expect from a large mass of rotating equipment.

この自己潤滑剤，ガラス含浸・黒鉛は，回転装置が大質量であることから想定される荷重および温度の過渡現象を**うまく処理できる**に違いない．

The essence of structured programming is to be able to use a general routine, for instance sorting and ordering, or one for reading in a piece of data, in any situation and thus reduce the work of solving a particular problem to **coping with** the

構成のはっきりしたプログラミングのエッセンスは，たとえば分類および順序づけ，あるいはまたデータの１部を読み込むのに，どんな状況でも汎用ルーチンを使うことができるという

difficulties that are unique to it.	ことで，特殊問題を解くとき，特有のむずかしい点を首尾よく**処理**できるようにすることだ．

deal with : to take action about or be what is needed by (a problem, etc.), to discuss in a book or speech etc.	**処理する，処置する，（問題として）取上げる**

Projecting rivet heads can best **be dealt with** by removing the shoes and countersinking the rivet heads.	突出たリベット頭は，シューを取外してリベット頭を皿取りすることによって，最も上手に**処理する**ことができる．
The modern thin-shell type of bearing which consists of a steel strip coated with white metal by a special process can not **be dealt with** by orthodox scrape-in methods.	最近の薄いシェル形軸受は，特殊な処理方法によってホワイト・メタルで被覆した薄鋼板からなり，従来のキサゲ法では**処置**できない．
Induced thrust loads can **be dealt with** the thrust ring with less friction and wear and without seizure.	誘起したスラスト荷重は，摩擦および摩耗が少なく，かつ焼付きのない，スラスト・リングで**処理**できる．
The results presented in this paper **deal with** the rolling contact fatigue life of tapered roller bearings obtained with six rear axle lubricants.	この論文に示した試験結果は，6つの後車軸潤滑剤を使って得た，円錐コロ軸受の転り接触疲労寿命を**扱った**ものである．

overcome : to win a victory over : to find a way of dealing with (a problem etc.)	**克服する，処理方法がわかる**

These problem **have been overcome** by a special variety of the double-pouring method.

これらの問題は，二重注入法のうちのある特殊な方法で，すでにうまく**処理済み**である．

process：to deal with a series of action or operation in making or manufacturing or achieving something	**処理する**

Process and arrange the data mathematically.

データを数理的に**処理し**て整頓する．

The unit can **process** up to 1,200 parts per minute with a 256×404 pixel field. It is designnned for high-volume assembly lines.

この装置は，256×404のピクセル（画素）フィールドを使って，毎分1,200個まで**処理する**ことができる．これは大量の組立ライン用に設計されたものである．

The cast-iron parts **are processed** on an 11 station.

この鋳鉄部品は，11ステーションで**処理する**．

Processed information is set to the protected memory.

処理された情報は，プロテクト・メモリにセットされる．

Process …… by computer.

……をコンピュータで**処理する**．

The washing fluid **is processed** through a multistage vacuum distillation unit which separate the coolant from the washing fluid.

洗浄液は，冷却剤を洗浄液から分離する多段真空蒸溜装置で**処理する**．

This phase of the program consisted of **processing** cutting tools with the highest rated coating and machining airframe

プログラムのこの段階は，最高級のコーティングで工具を**処理する**ことと，生産現場で航空

components under production environment.

機機体部品を切削することである.

The coating **has been processed** to meet MIL-L-46010.

このコーティングは，MIL-L-46010に合うように**処理した.**

B. Inc. were having trouble in obtaining a suitable finish on certain copper alloy parts. After they are machined, parts frequently had to **be processed** to remove stain or tarnish.

B社は，ある銅合金部品を適当な仕上りにすることに苦労していた．部品を切削してから，**錆や曇りを除去するために処理しなければならないことが**再三であった.

The basic components of an intelligent machine include a mechanical system which performs the actual machining **process,** and a control network which operates the mechanism.

知能機械の基本要素の中には，実際に切削**処理をする**機械システム，およびそのメカニズム(機構）を動かす制御ネットワークがある.

Since **processing is done** in an open container, parts are easy to inspect and it is possible to see whether the desired results are achieved.

開放された容器の中で**処理するので，部品は検査が楽で，**また所望の結果が得られたかどうか見ることができる.

〈用　語　例〉

process (in a data processing system)　処理(データ処理システムの)
process annealing　中間焼なまし
process capability chart　工程能力図
process control　プロセス制御
process specification　作業標準
processed vidio　処理済みビデオ
processing aid　加工助剤
processor　処理装置

treat: to deal with (a person or thing) in a certain way to present or deal with (a subject)　**処理する，扱う**

Remember to **treat** all bearings care-

軸受はすべて注意深く**取扱う**

fully. Arrange for a suitable stop or support for the shaft, otherwise the bearings may be damaged by the dismounting forces normally occurring during the operation.

ことを忘れずに．軸に適した止め具あるいは支えを用意しておく．そうしないと取外し作業中に生じる力で，軸受を損傷することがある．

It is not always successful on a painted finish, so before **treating** the whole car it should be tried on an inconspicuous part.

塗装の仕上げが必ずうまくゆくという訳ではない．したがって，車全体を**処理する**前に，見えにくい部分にそれを試しにやってみることである．

Recently, Soviet industry has introduced a method for finishing steel gears that employs plastic surface deformation (i.c., burnishing). Gears of up to 62 Rockwell C hardness **are** so **treated**, and the burnishing tools are gears, with the body made of steel **treated** to 35-40 Rockwell C and inserted, brazed carbide teeth.

最近，ソビエト工業界は，塑性表面変形（すなわち，バニシング）を用いて，鋼の歯車を仕上げる方法を採用している．硬さ62RC までの歯車が，そう**処理されている**．またバニシング・ツールは歯車で，RC35～40に**処理した**鋼で作ったボディと，はめ込み・ろう付けした超硬の歯が付いたものである．

This book **treats** of an important subject.

この本は，重要なテーマを**取上げ**ている．

Water **is treated** as an imcompressive fluid.

水を，非圧縮液体として**取扱う**．

The way a system is updated and how data is verified often **are treated** too lightly.

システム更新およびデータ検証の方法を，軽々しく**取扱い**すぎることが少なくない．

〈用　語　例〉

heat treatment　熱処理
pretreatment　前処理
phosphate treatment　燐酸処理
quality of treated water　処理水質
treated (grinding) wheel　処理砥石

調べる（観察，看視，調査，走査，検査）

assay: testing of metals, especially those used for coin or bullion, for quality.	調べる（試金）

The actual "hall" mark shows where the metal **was assayed** (the leopard stands for London), and a date stamp (in script above, F indicates 1980).

この実際の刻印は，この金属が**調べられた**場所（豹はロンドンを表わす）と日付刻印（上記手書きの，Fは1980年を示す）を示している．

delve: to dig : to search deeply for information	調べる，掘り下げる

It is necessary to **delve** deeply **into** this complex problem.

この複雑な問題を深く**調べる**ことが必要である．

examine: to look at closely : to look at in order to learn about or from	調べる，よく見る（検査，検出，点検，検討）

Some of the bearings **have been examined** when wheels have been removed.

ホイールを取外したときに，いくつかの軸受を**調べた**．

While the plug **is being examined** the spark gap should be checked.

プラグを**調べる**時に，スパークギャップを確かめること．

The cams should **be** carefully **examined for** wear.

カムは，摩耗しているかよく**調べる**こと．

Carefully inspect the edges of the valves and the seating to see whether they are pitted, and **examine** the stem to see whether they are straight or worn.

弁の縁および座が，ピッチングを生じているかどうか見てよく検査し，かつ弁棒が曲がっていないか，また摩耗していないかどうかを見て**調べる**．

The balls and raceways **were examined** by optical microphotography.	玉および軌道を，光学顕微鏡写真で**調べた.**
The disk surface **was examined** with XPS for a lubricant containing dibenzyl disulfide (DBDS) as an antiwear additive.	ディスク表面を，耐摩耗剤として二硫化ヂベンジルを含む潤滑剤について，XPS で**調べた.**
……**make** the **examination of** ……	……について**調べる.**
Competent **examination** for mechanical causes **be made**.	機械的原因について適切な**調査をすること.**
Metallurgical **examination** of the sectioned cutters did not reveal a change in micro-structure.	切断したカッタの冶金学的**調査**では，顕微鏡組織に変化は見られなかった.
Typical wear scar traces are shown in Fig. 5 . **Examination** of Fig. 5 shows that the wear scar indentation is very nearly parabolic in form.	摩耗痕跡の例を図 5 に示す．図 5 の**調査**では，摩耗痕のへこみは形が放物線にきわめて近いことを示している.

explore : to examine by touch　　**調べる**

It presents a menu showing all the parameters that **can be explored** for either surface or roundness measurement.	表面か真円度測定のいずれかを**調べる**，すべてのパラメータを示すメニューが，表示される.
The potential for such geometries has not **been** fully **explored** by tool manufacturers.	ツール・メーカーは，そういう形状寸法の可能性について，完全に**調べ**尽してはいない.

inspect : to look upon, to view closely and critically
: to examine (a thing) carefully and critically, especially looking for flaws　　**調べる，詳しく見る(検査)**

Carefully **inspect** the components close to the bearing location. Remove burrs and clean shafts and shoulders.

軸受位置に近接している構成部品を注意して**検査する**．バリを取除き，軸および肩部をきれいにする．

Although rolling bearings are robust mechanical components which give long service time it is, however, wise to **inspect** them now and then.

転り軸受は，長時間使用できる頑丈な機械部品ではあるが，それでもときどき**検査**したほうがよい．

The clutch linings should **be** carefully **inspected** when deciding whether renewal is required.

クラッチライニング更新の要否を決めるときには，注意して**検査すること**．

Sealed or shielded bearings should not be washed on any account ; for obvious reasons they cannot **be inspected** either.

シールあるいはシールド軸受は，どんなことがあっても洗わないこと；理由は明らかで，どちらも**検査**できないからである．

This may **be inspected** on the actual wiper by removing the outer cover.

この点については，外側カバーを取外すことによって，ワイパーそのものを**検査**できる．

Before removing a tube which has become slowly deflated **inspect** the valve for leakage.

空気が徐々に抜けるようになったチューブを取外す前に，バルブの漏れを**検査する**．

Before attempting to remove the pinions they should **be** carefully **inspected** for wear and backlash, and it should be verified that timing marks are provided on the teeth.

ピニオンを取外そうとする前に，摩耗およびバックラッシュを注意して**検査すること**．そして歯にタイミングマークがついていることも確かめること．

Visually inspect parts for obvious defects that would render the part unserviceable.

部品が使えなくなるような，はっきりした欠点がないか部品を**目視検査する**．

Each lot **is** 100% **inspected** for quality both externally and internally.	ロットごとに内・外両方の品質を100%**検査する.**
Inspect the slide to see that it moves freely and its rollers do not bind.	スライドが自由に動き，ローラもひっかからないか，見て**検査する.**
Inspect all rollers to be sure that they are evenly covered with a thin film if ink.	インクをつけたら，間違いなくすべてのローラが薄い膜で均等に覆われているか**検査する.**
Dye penetrant **inspect** the fittings to determine if there are any indications of cracks, specifically in the area of the four corners as indicated below.	亀裂の形跡があるか見極めるため（特に下に示す四隅の部分に），金具を染色浸透剤で**検査する.**
HSS milling cutters **were inspected** to verify the tool gemetry.	高速度鋼製フライスの工具形状寸法を確認**検査した.**
Turned parts and shaft can **be inspected** 20 times faster than usual with the gage shown.	旋削した部品および軸は，図示のゲージで，普通よりも20倍早く**検査**できる.
The fan blades have been cracked when **inspected** by magnaflux at the 1,200 hour inspection interval.	1,200時間点検において，マグナフラックスで**検査したときに**，ファンブレードが亀裂を生じていた.
Inspect yoke assembly by magnetic particle method, with specific emphasis to threaded portion of yoke spindle.	ヨークスピンドルのネジ部に特に重点をおいて，ヨークアセンブリを磁粉探傷法で**検査する.**
Wash the exposed bearing where it is possible to **carry out inspection** without dismounting.	取外さずに**検査する**ことができる場合，露出した軸受を洗浄する.

In addition **an inspection** should **be performed** any time exhaust fumes are detected in the cabin.

さらに，部屋に排気の臭気が検知されたときは，どんなときでも，**検査する**.

100 HOUR **INSPECTION**.
Accomplish each 100 hour interval.

100時間**点検**.
100時間ごとに行なう.

Disassemble hub and blade only to extent required to **accomplish inspection** and parts replacement.

検査および部品交換の必要がある範囲だけ，ハブおよびブレードを分解する.

Inspection is applied at this stage.

この段階で，**検査する**.

……**is subjected to inspection**.

……を**検査する**.

Submit the parts to **inspection**.

部品を**検査**に提出すること.

……present the automobile for **inspection**.

自動車を**検査**に出す.

Inspection of axle (after test) reveal no unusual wear as measured by dimensional preload and stack height of bearing.

車軸の**検査**（試験後）で，軸受寸法による予圧および組み幅で測ったが，異常な摩耗はまったく見られなかった.

Replace levers if found defective by fluorescent penetrant **inspection**.

もし，螢光浸透剤**検査**（螢光探傷）で欠陥が発見されたら，レバーを交換する.

Check the lubricant. Impurities of various kinds can usually be felt if a little of the lubricant is rubbed between the fingers ; or a thin layer may be spread on the back of the hand to **inspection** against the light.

潤滑剤をチェックする．いろいろな種類の不純物は，少量の潤滑剤を指でこすれば，通常，感じでわかる；光にかざして**検査するのに**，手の甲に薄く伸ばしてもよい.

The drive shaft, **inspection** and lubrication frequency is each 100 hours.

駆動軸の**検査**および給油頻度は，100時間ごとである．

Conventional visual **inspection** method can not gage the extent of a seam or crack in a steel bar.

普通の目視**検査**方法では，棒鋼の亀裂や傷の範囲は測れない．

〈用　語　例〉

acceptance inspection　受入検査
delivery inspection, outgoing inspection　出荷検査
final inspection　最終検査
in-process inspection, inspection between process　中間検査
inspection gauge　検査ゲージ
inspection hole　検査孔
item-by-item sequential inspection　各個逐次抜取り検査
normal inspection　なみ検査
purchasing inspection　購入検査
reduced inspection　ゆるい検査
tightened inspection　きつい検査

monitor : to keep watch over
: to record or test or control the working of　**モニターする，監視する**

The quality of oil **is monitored** automatically.

油の品質を自動**モニターする**．

Cutting forces **were monitored** at one minute interval by a time selector switch in the system 8000 data logger.

切削力を，システム8000データロガーのタイムセレクタ・スイッチで1分間隔で**モニターした**．

Monitor lights can be clearly observed from a distance when the machine is in operation.

監視灯は，機械が動いているとき，離れたところからはっきり観ることができる．

〈用　語　例〉

Auto Monitor Program　オートモニター・プログラム

observe : to see and notice : to watch carefully	**調べる（よく見る，観察）**

Observe that it is not necessary to dismantle anything to apply the necessary lubricating oil.

必要な潤滑油を供給するのに，取外す必要のあるものが何かないか，**よくみる．**

Possible reasons for the differing chemistries **observed** in the field tests of Oil A with those **observed on** Oil B …….

実地試験で**観察した**油Aの化学性状が，油Bで**観察した**性状と違う理由で考えられることは，…….

…… **observed** the soap fibers of the grease by means of an electron microscope and obtained some new knowledges.

……電子顕微鏡でグリースの石鹸繊維を**調べて**みて，若干新しいことがわかった．

Make observations on seal leakage.

シールの漏れを**よく調べる．**

After 300, 1,000 and 25,000 hours, the bearings are dismantled for **observation** of the components and the lubricant ; once the **observation is accomplished**, the bearings are assembled again, without replacing the lubricant.

300，1,000および25,000時間後に，構成部品および潤滑剤を**よく観るために**軸受を取外す；**調べが済んだら**，潤滑剤を変えないで，軸受をふたたび取付ける．

〈用　語　例〉

observation tank　　検油タンク

probe : a device for exploring an otherwise inaccessible place or object etc. : to make a penetrating investigation of	**調べる（探査，探測，探針）**

Operating high above the earth's atmosphere, the telescope will **probe** seven times deeper into space, detect objects 50 times fainter, and view them with ten times

地球の大気圏よりも高いところで操作すれば，望遠鏡は地上の天文台でできるよりも宇宙を7倍深く**探査し**，50倍も微小な

better clarity than is possible with ground-based observatories.

物体を検出し，かつ物体を10倍はっきり見られる．

〈関連用語〉

scan　調べる，走査（光分析の）スキャン
scanner　走査装置
scanning disk　走査板
scansion　精査
scanning line　走査線
scanning search　スキャニング・サーチ
scanning speed, scanning rate　走査速度

read: to study or discover by reading　**調べる（点検）**

Read a meter.

メータを**調べる**．

〈用　語　例〉

read group total, reset group total　グループ合計点検
read total key　点検カギ

scrutinize: to look at or examine carefully　**詳しく調べる（精査，吟味）**

Refrigerant fluid or oil leakage is closely **scrutinized**.

冷凍液あるいは油の漏れを，**綿密に調べる**．

see: to look at for information.　**調べる**

See occasionally that there is no oil leak from the plug.

折りにふれて，プラグから油が漏れていないか**調べる**．

〈用　語　例〉

market survey　市場調査

study: to give one's attention to acquiring knowledge of (a subject)
: to examine attentively
調べる（研究）

A **study** by M. as well as in our laboratories suggested that most of the heat generated in a new axle at light speed is due to the wearing-in of bearing surface.

M氏とわれわれの研究所の**調べ**では，中速での新しいアクスルの発生熱の多くは，軸受表面の馴染みによるものと思われる．

survey: to look at and take a general view of : to examine the condition of (a building etc).	**調べる（実地踏査，測量，検査）**

carefully **survey** the recent literature concerning…….

……に関する最近の文献を注意して**調査する**.

During the last ten years, we **have surveyed** the industry in search of useful test procedure for the evaluation of this phenomenon.

この10年間，この現象を評価するのに役立つ試験方法を探し求めて，この産業界を**調査した**.

……**make a** small **survey of**…….

……について小規模の**調査をする**.

test: a critical examination or evaluation of the qualities or abilities etc. of a person or thing; to subject to test	**調べる（検定，検査，試験）**

These diagnostics **test** the microprocessor system, the disk storage unit, all inputs and outputs of the computer, the alpha-numeric display and the alpha-numeric keyboard.

これらの診断は，マイクロプロセッサシステム，ディスク記憶装置，コンピュータのすべての入力と出力，文字数字表示および英数字キーボードを**テストすることである**.

The same time-**tested** components we've used on other machines.

耐久**試験した**ものと同じ部品を，他の機械にも使っている.

The overload relays **are** individually factory-**tested** and calibrated.

オーバーロード・リレーは，1つ1つ工場で**試験し**，補正してある.

A turbine engine mainshaft-type, an-

タービンエンジン主軸用の

gular contact, thrust bearing **was** rig **tested** to 1.5 million DN. Cage motion in both the radial and axial planes, total heat generation rates and bearing torque were measured and compared to the results of two computer models. A relation between bearing power loss and cage kinematics was observed.

アンギュラコンタクト，スラスト玉軸受を，DN1.5×10^6まで装着**試験した**．半径および軸方向の保持器の運動，全発熱量および軸受トルクを測って2つのコンピュータモデルの結果と比べ，軸受の動力損失と保持器の運動力学との関係を調べた．

The sampling method **was** field **tested** to determine the method precision.

その方法の精度を求めるために，サンプリング法を現場**試験した**．

It is best **tested** by means of a feeler guage.

すきみ計で**調べる**のが，一番いい．

Before and after the **test**, the bearings **were tested** individually for the vibration level on a Labo instrument at 1,000 rpm and axial load of 4N.

試験の前後に，研究所の装置で，個々の軸受を1,000rpm，軸方向荷重4Nで振動レベルを**試験した**．

D.H. of the D Company, in 1904, **did** a series of **tests** with emery and corundum grinding wheels which proved that corundum was far superior.

1904年，D社のH氏は，エメリーおよびコランダム研削砥石を使って一連の**試験をし**，コランダムが遙かに優れていることを実証した．

The **test is carried out** on this instrument.

この装置で**試験する**．

Particularly, steel manufacturers must **perform** some kind of **test** to check the machining properites of their products.

特に製鋼メーカーは，その製品の切削性を確かめるために，ある種の**試験をし**なければならない．

Make a Rockwell **test** on three specimens using the correct penetrator, major load, and scale.

正確なペネトレータ，大きな荷重とスケールを使って3つの試料をロックウェル**試験する.**

All **tests were conducted** in the M. tester.

試験はすべて，M試験機で**行なった.**

Laboratory **tests were conducted** at low speed (9 rpm) and at high speed (3,000) on a 228 mm diameter journal bearing.

研究室の**試験は**，直径228mmのジャーナル軸受について，低速（9 rpm）と高速（3,000）で**行なった.**

The **tests were run** on the low speed test machine.

試験は，低速試験機で**行なった.**

The **tests were run** on diamond grinding wheel only.

試験は，ダイヤモンド砥石だけについて行なった.

Put products **to** the **test**.

製品はこの**テストをする.**

Two such bearings **were** built and **put on** life **test**. This report describes their construction, and their condition after 13, 000 /h life test.

こういう軸受を2個製作して，寿命**試験した.** ここに，その構造および13,000時間寿命試験後の状態をレポートする.

The bearings **were subjected to** centrifugation **tests** to determine their oil retention characteristics.

ベアリングはそのオイル保持特性を決めるための遠心**テストをした.**

Each **test** gave a different value of tool life.

工具寿命は，**試験**ごとに違う値になった.

One **test** produced almost the same wear behavior as the standard spline, while the other **test** produced appreciably better performance as shown.

1つの**試験**は，摩耗の様相が標準スプラインとほとんど同じになったが，別の**試験**では図示のようにかなり良い性能であった．

Laboratory and field **testings** permit the evaluation of the effects of design, material, and production variables on performance of the part under controlled conditions.

研究室および実用**試験**によって，設計，材料および製作変数が管理条件下で部品性能に及ぼす影響を評価できる．

The usual laboratory wear and extreme pressure **tests** have found some limited use.

研究室における通常の摩耗および極圧**試験**で，使用上若干の制約のあることがわかった．

Tapping **tests** on metals like SAE 3140…… fail to show differences between products.

SAE3140のような金属のタップ加工**試験**では，製品間の差を見付けることはできなかった．

The second-time **test** exhibited greater scatter in wear behavior due to the varying wear obtained in the first-time **test**.

第2回目の**試験**では，第1回の**試験**で得られた摩耗がいろいろあったため，摩耗の様相のバラツキはより大きく現われた．

……These **tests** clearly demonstrated improvement in drill life owing to the application of a coolant.

……これらの**試験**は，冷却剤を与えることによりドリル寿命が向上することを明瞭に示した．

The **tests** confirm the unsuitability of the lubrication systems for this particular application.

この**試験**で，この特別な用途に対する潤滑システムが適切でないことがはっきりした．

Tests established that 5% barium hydroxide in water gave nearly double the tool life of other coolants tested.

水酸化バリウムを5％含んだ水の場合，他の冷却剤で試験した場合より工具の寿命が2倍近くなることが，**試験**によって立証された．

The tapping **test** has proven its value as

このタップ加工**試験**は，切削

an effective screening method to predict performance of cutting oils, soluble oil emulsion and, more recently, aqueous coolants.

油，水溶性油剤およびごく最近の水性冷却剤の性能を予測する有効な篩分け法に値することを，立証した．

A total 48 end milling cutters were used in screening **test**.

全部で48のエンドミルカッタを，ふるい分け**試験**に使用した．

To obtain the experimental values of torsion, heat generation and cage speed, a series of **test** runs **were conducted** where the bearing was subjected to the predetermined sets of load and speed conditions.

捻り，熱の発生および保持器速度の実験値を得るために，一連の**試験**運転**を行なった**．この場合，軸受にはあらかじめ決めた組合せ荷重と，速度条件を与えた．

It would have to be verified by further **testing**.

さらに**試験**して，確かめなければならないであろう．

〈用　語　例〉

distribution-free test　分布によらない検定
nondestructive test　非破壊検査
program test　プログラムテスト
rotation test, speed test　回転試験
visual test　目視検査

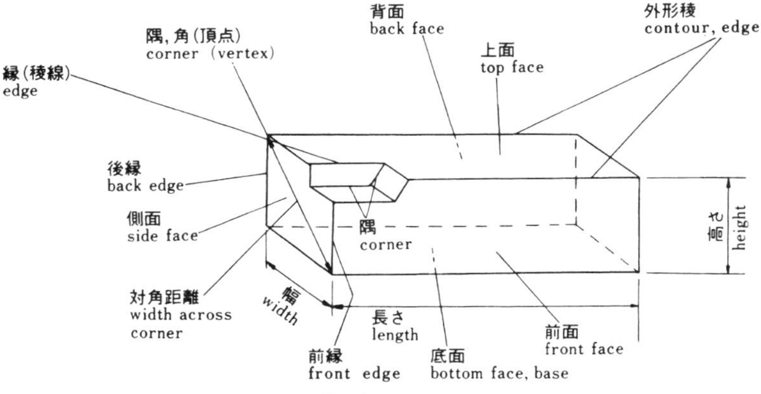

面，縁，角
face, edge, corner

する，やり遂げる（実行，完遂）

accomplish : to succeed in doing, to fulfil : to work toward the end, realize a project	する，やり遂げる，完遂する

A standard CNC table mechanism can be used to **accomplish** this positioning.	この標準 CNC テーブル機構は，この位置決めを**する**ために使うことができる．
Drying, curing, and posturing **were accomplished** after removal of the wound tubes from the mandrel.	巻いたチューブを心棒から取外したあと，乾燥，熟成，整形を**した**．
Positioning **is accomplished** <u>by</u> hardened, precision indexplates and pins.	位置決めは，焼入れた精密割出し板とピン<u>で</u>**する**．
Deep hole drilling of small holes generally requires slower speeds, and **is** usually **accomplished** <u>by</u> gradually increasing the hole diameter and depth, using the most rigid drills possible.	小さい穴の深穴明けは，できるだけ剛いドリルを使って，穴の直径および深さを次第に大きくすること<u>によって</u>**行なう**．
Work holding **is accomplished** either <u>by</u> using a machine vise bolted to the worktable or <u>by</u> direct bolting of the workpiece onto the worktable using the T slots provided.	工作物の保持は，ワークテーブルにボルト止めした機械バイスを使うか，既設のT溝を使って工作物をテーブルに直接ボルト止めするかの<u>方法で</u>**する**．
Production of screw thread can **be accomplished** <u>by</u> the use of taps and dies.	タップとダイスを使うこと<u>によって</u>，ネジを製作**する**ことができる．
Where it is desirable to provide means	潤滑剤を更新する手段を講じ

of renewing the lubricant, this may **be accomplished** either by means of a grease cup screwed into the closure cap or a pressure gun fitting located in the position shown in Fig. 16.

ることが望ましい場合，これは閉じ蓋にねじ込んだグリース・カップ，または図16に示した姿勢に位置決めした圧力ガン金具のいずれかの方法で**行なう**ことができる.

The one rpm incremental speed changes **are accomplished** through an all-gear drive.

1回転単位の速度変換は，全歯車駆動装置によって**行なう**.

Vane-to-housing sealing **is accomplished** through machining rather than elastomeric seals.

翼とハウジングの密封は，エラストマのシールではなく，機械加工で**する**.

When required forming operations can **be accomplished** with 3 tool positions and cutting-off,…….

必要とする成形作業を，工具位置3条件および突切りで**する**ことができるときは……

Starting and stopping under load can **be** easily **accomplished.**

負荷状態で，楽に始動，停止を**する**ことができる.

A diagnostic package makes it a simple matter to find and correct a malfunction. Complete precision parts processing from a single station setup **is accomplished** in a fraction of the time required for multiple machine processing.

診断パッケージによって，機能不良を発見して直すことは簡単である．1つのステーションでの段取りによって精密部品を完全に処理することが，複数の機械処理による時間のほんの一部で**すむ**.

Staking must **be accomplished** within 30 minutes after application of adhesive to mating surfaces of parts. Never stake bearings after curing of adhesive.

部品の合わせ面に接着剤を塗ってから30分以内にかしめを**完了**しなければならない…….

apply: to put into effect　　～をする

Inspection **is applied** at the appropriate stage.

適当な段階で，検査**をする．**

effect: to bring about, accomplish　　（～を）する，してしまう

A circular punch, located in the upper die, and working in conjunction with the blanking die, **effects** the disk-cutting operation.

上型にあり，打抜きダイスとともに働く丸いパンチが，円板を切る作業**をする．**

Once the program is in memory, a manual data input device can be utilized to **effect** any changes.

一度プログラムが記憶されると，どんな変更を**する**にも，マニアルデータ入力装置が使える．

In sawing the removal of metal **is effected** by a series of saw teeth. With power-driven band saws and circular saws, cutting can be performed in a continuous operation.

鋸で切るとき，金属の切削は，一連の鋸歯で**行なわれる．**動力駆動の帯鋸および丸鋸を使うと，連続作業で切削をすることができる．

In the event of the air-bag puncturing, a repair can easily **be effected** with a tyre patch.

空気袋がパンクしたような場合には，タイヤパッチで容易に修理を**する**ことができる．

Tool changes can **be effected** in less than 90 minutes.

ツール交換は，90分以内で**完了**できる．

implement: to put into effect　　する，実行する，具体化する

As often happens when a radical process change **is implemented**, problems were

急激な工程変更を**する**ときよく起こるように，問題にぶつか

encountered.

った.

When **implemented** by the machine and tool designers, should assure the continued successful broaching of engine components.

機械および工具設計者が**する**ときには，エンジン部品の連続ブローチ加工がうまくできることを確かめること.

(～を) する (実行，実施)

carry out: to put into practice, to accompish	(～を) する，実行する

Production **is carried out** automatically with an automated cam-controlled transfer system feeding the parts into the double tooling as shown in Figure 2.

図2に示すように，部品をダブル・ツーリングに送り込む自動カム制御トランスファ・システムを用いて，生産は自動的に**行なわれる**.

A neat repair of a frayed floor carpet may **be carried out** with a patch of leather of the same colour as the carpet.

ほつれた床カーペットをきれいに修理するには，カーペットと同じ色のレザーパッチを使って**する**ことができる.

Pocket milling. can **be carried out** with guaranteed accuracy in all 3 axes.

フライスによるポケット削りは，3軸すべて保証精度で**する**ことができる.

A facing operation can **be carried out** by using a special tool holder.

特殊なツールホルダを使うことによって，面削り作業を**する**ことができる.

Two experiments **were carried out** at room temperature to assess the effect of water and steam contamination on the performance of the bearing.

軸受の性能に及ぼす水および蒸気汚損の影響を評価するため，室温で2つの実験を**した**.

In some cases it may be necessary to remove the engine from the frame before work can **be carried out** on the clutch.

クラッチに作業するには，しばしばその前にエンジンをフレームから取外すことが必要になる場合がある．

conduct: to manage or direct (business or negotiation etc. or an experiment)	（工夫して）する，処理する

The controller **is conducting** its error check, ····

制御装置がその誤りのチェックをしている，……．

The technique used to **conduct** these analyses the Spectrometric Oil Analysis Program (SOAP)....

これらの解析**をする**のに使われる技法のスペクトロメトリック油分析プログラム(SOAP)は……．

The field trial **was conducted** at a M maintenance shop.

現場での試行は，M保全工場で**行なった**．

For many years, tests **were conducted by** various companies in order to find a grease that best resists separation when subjected to high centrifugal forces.

高遠心力を受けたとき，最も分離しにくいグリースを見付けるため，いろいろな会社が長い年月にわたり**試験をした**．

In certain experimental rig tests, **conducted** on 124-mm bore bearings at speeds to 3.0 MDN, this has **been done** on an optimized design.

内径124mmの軸受を，3×10^6DNの速度で**実施した**実験装置の試験は，最適設計のものについて**行なわれた**．

All the friction and wear tests **were conducted** using a Falex LFW-1 machine.

摩擦および摩耗試験はすべて，F機を使って**行なった**．

Research in this area **is being conducted** at the Production Engineering Institute of the Aachen University.

この分野の研究は，アーヘン大学の生産技術研究所で**行なわれている**．

Many **tests were conducted** to measure the friction coefficients in couplings, and to find means to reduce them.

カップリングの摩擦係数を測り，それを減らす手段を見付けるため，数多くの試験**をした．**

do : to perform, to carry out, to fulfil or complete (a work, duty etc.)

（～を）する

While it has size and power limitations, this type of press **does** particular jobs efficiently.

寸法および出力に制限があるが，このタイプのプレスは特殊な仕事を効率良く**する．**

The 2A has a math processor option, which allows engineers to **do** calculation in real time and read the results on the hand-held terminal.

2Aは，オプションで数理プロセッサがあり，技術者はハンドヘルド端末機を使い実時間で**計算して，結果を読むことができる．**

When removing a washer, all that he **has to do** is to measure its thickness and **make** a simple calculation.

ワッシャを取外すときに，**しなければならない**ことは，その厚さを測って，簡単な計算を**することである．**

This operation should only **be done** by experts.

この作業は，専門家だけが**すること．**

Screw cutting for a limited number of components can **be done** cheaply and simply by hand with a screw tap or a screw die.

限られた数の部品のネジ切りは，タップまたはダイスを使って，手で安価かつ簡単に**することができる．**

This can **be done** very simply as shown in Fig. 146, by connecting a bulb-holder between the two outer terminals of the two-way switch.

これは，図146に示すように真空管保持具を２路スイッチの２つの外側端子の間に接続することによって，ごく簡単に**することができる．**

Much lathe work **is done** in a chuck and requires considerable facing and some center drilling.

旋盤作業の多くは，チャックで**する**が，かなりの面削りおよびセンタ穴明けも必要である.

The Model H has a single-lever control which performs indexing and spindle-locking. Indexing and locking **are done** in ratchet fashion.

H 型には，割出しおよび軸の固定を**する**レバーコントローラがある．割出しおよび固定は，ラチェット方式で**する**.

Mechanical drilling of small holes can often **be done** on conventional machine tools such as drill presses, lathes, jig boring machines, machining and turning centers, and transfer machines.

小さい穴の機械的穴明けは，ボール盤，旋盤，治具中ぐり盤，マシニングセンタ，ターニングセンタおよびトランスファ・マシンのような普通の工作機械で**する**ことが多い.

In the last two decades of the 19th century intensive research **was done** on the processes of metal cutting.

19世紀末の20年間，金属切削のプロセスについて精力的に研究が**行なわれた**.

When lubrication **has been overdone** so that the brake-linings have become covered with grease or oil, the only thing is to remove the shoes and thoroughly wash them in petrol, rubbing them over with a wire brush.

ブレーキ・ライニングがグリースや油で覆われるようになるほど潤滑**しすぎてしまった**ときには，シューを取外して，ガソリン中で完全に洗い，ワイヤブラシでその上をこする以外に方法はない.

execute: to carry out, to put (a plan etc.) into effect

する，実行する

It is possible to **execute** very close delivery schedule of both incoming material and outgoing finished parts.

入ってくる材料と，出てゆく仕上げ部品の両方をきわめて綿密な受け渡しスケジュールで**実行する**ことができる.

Assembly requirements analysis indicated that the majority of assembly tasks **are executed** with a vertical Z-motion.

組立に必要な条件を解析して，大部分の組立作業は上下のZ運動で**実行できる**ことがわかった．

Since accurate machining **is executed** as the program is being prepared, the Word Player is ideal for the production of single workpieces.

準備したプログラム通りに正確に機械加工**する**から，ワードプレーヤは単品工作物の生産に理想的である．

make: to perform (an action etc.)	**する**

Feed the compound in 0.005 inch and reset the cross feed dial to zero. **Make** the second cut.

複式刃物台を0.005インチ送り込み，前後送りダイヤルをゼロにセットし直し，２回目の切削を**する**．

On a lathe, long shafts tend to vibrate when cuts **are made** leaving chatter marks.

旋盤で長い軸を切削**する**ときは振動しがちで，びびりマークが残る．

The grinding of high-chromium iron rolls can **be made** with the same equipment and procedures that are used for grain iron rolls.

グレーン鉄ロールに用いたのと同じ装置および作業要領で，高Cr鉄ロール研削を**する**ことができる．

The system utilizes 24 nonrotating weld guns that **make** six welds on each baffle.

このシステムは，各そらせ板に６ヵ所の溶接を**する**24の非回転溶接ガンを使っている．

Now you can **make** highly precise measurements more rapidly to minimize downtime of high productivity machine tools with the Leitz family of CNC Coordinate Measuring Machines.

Leitz の CNC 座標測定機で，高生産工作機械のダウン・タイムが最も少なくなるよう，より迅速に，高精度な測定を**する**ことができる．

Regarding the question on leakage rates,

漏れ速さの問題に関しては，

no sophisticated measurement of flow **was made** during the test, only qualitative observations with the naked eye.

試験中は複雑な流れ測定はまったく**行なわず**，裸眼での定性的観察だけである．

Running at the moderate speed of 2,000 rpm the piston, of course, **makes** 4,000 strokes per minute and in doing so travels about 1,120 ft.

2,000rpmの中速で運動しているピストンは，もちろん1分間に4,000ストローク**し**，それは約1,120フィート移動することになる．

Current is led from the dynamo by means of carbon blocks or brushes which **make** rubbing contact with the commutator.

コミュテータとこすり接触**する**カーボン・ブロック，すなわちブラシによって，電流はダイナモから導かれる．

……; in the third position both resistances are short-circuited, connection **being made** directly to the end of the field winding.

……；第3の位置で，両方の抵抗が短絡し，フィールド巻線の端に直接連結**する**．

Having made these preparations, the engine should be run until it is nicely warm, as joints are broken more easily and nuts are easier to remove when warm.

これらの準備を**しておいて**，エンジンをほどよく暖まるまで運転すること．暖かければジョイントは楽に解け，ナットは取外しやすい．

In that case, new provision will have to **be made** for carrying the tools, but this is not usually a difficult matter.

そんな場合には，ツールを取付けるための新しい段取りを**しなければならない**が，これは普通むずかしいことではない．

The M portable machine could be adapted to the job, providing certain design changes **were made**.

特定の設計変更を**すれば**，M可搬機械はこの仕事に適応させることができる．

On cars not fitted with a stop-light the

停止灯を取付けていない車で

addition can **be made** quite easily by attaching the switch to the underside of the floor-board and hooking the end of the spring through a hole made in the brake pedal.

は，床板の下にスイッチを取付け，ブレーキ・ペダルに作った孔を通してバネの端をひっかけることにより，ごく簡単に追加**することができる．**

Divisions can **be made** that are not possible with a standard index plate.

標準の割出し板ではできない分割を**する**ことができる．

Practical necessity sometimes requires that we **make** this kind of classification.

実際上の必要性から，この種の分類を**する**ことが時に必要になる．

Not only should a note **be made** of the performance, but fuel consumption should also be taken into account, for it can be changed very considerably by quite small alterations in jet sizes.

ジェットの寸法をほんのわずか変えても燃料消費量はかなり変わることがあるので，性能に注意**する**だけでなく，燃料消費量も考慮すること．

Increment infeeds and cycle selection **are made** by push button controls.

インクリメント・インフィードおよびサイクルの選定は，押ボタン・コントローラで**する．**

……**make** every effort to eliminate errors.

誤りをなくすように，あらゆる努力を**する．**

Attempts are **being made** everywhere to standardize in order to **make** them portable —— to **make** it possible to run on different operating systems and machines.

持ち運べるように**する**ため——異なる操作システムおよび機械で運転できるように**する**ため——標準化する試みが至るところで**なされている．**

Two years ago, a study **was made** of the company's total belt operations.

2年前，会社のすべてのベルト作業について研究**した．**

A further test could **be made** by opening

エンジンがゆっくり作動して

the throttle suddenly while the engine is running slowly

いるとき，急にスロットルを開くことによって，さらに**試験をする**ことができる．

Sources of Leakage.——When it is found that too much oil is being used a careful **check** of the engine should **be made**.

漏れの源——油が過剰に使われていることがわかったときは，エンジンを注意深く**調べる**こと．

The next step was to **make** further observation or experiments to determine whether or not these other conditions were present.

つぎのステップは，これら他の条件があるかないかを判定するため，さらに観察あるいは実験**する**ことであった．

In order to test his hypothesis, Dr. A had first to **make** certain assumptions.

仮説を確かめるために，A博士は最初に特定の仮定を**した**．

No assessment has **been made** of the effects of the additive on the corrosion and wear characteristics of the engine or on combustion chamber deposits.

この添加剤がエンジンの腐蝕，摩耗特性および燃焼室の堆積物に及ぼす影響については，まったく評価**しなかった**．

Make sure that ……

……を**確かめる**（確保する）．

When writing even short and simple program in a computer language everyone **makes mistakes**.

コンピュータ言語で短かい単純なプログラムを書くときでも，誰もが**間違える**．

This point will more readily be understood by **making** reference to Fig. 44 which shows …….

この点は，……を示す図44を参照**する**ことによって，もっと簡単にわかる．

The gearing industry **made** significant gear design advances which necessitated the production of hobs with modified tooth forms.

歯切産業は歯車設計の顕著な進歩を**実現した**が，それには修整歯形ホブの製造が必要であった．

The tool that **made** the biggest contribution is the Valenite Centre-Dex with one screw-down square VC-55 insert with an 11° relief.

最も大きく寄与**した**ツールは，11°の逃げ面をもち，ネジ1本で締める四角のVC-55インサート付きのValenite Center-Dexである．

……to **make** more successful and systematic use of his computer.

もっとうまく組織的にコンピュータを活用**する**ために……

perform: to carry into effect, to accomplish, to do

する

At this point it momentarily stops and **performs no work**.

この点で瞬間的に停まり，**まったく作動しない．**

Called the Parts Maker, the CNC vertical-spindle, bed-type machine **performs** drilling, milling, boring, tapping, and precision profiling in three axes simultaneously.

Parts Makerという，このCNC立軸ベッド形機械は，同時3軸で穴明け，フライス削り，中ぐり，タップ立ておよび精密ならい削りを**する．**

The arm **performs** a range of motion within a sphere 16 inch diameter.

腕は，直径16インチの行動空間内で運動**する．**

These instruments are designed to **perform** one specific function, or can be combined to achieve a number of different functions.

装置は，1つの指定された機能を**果たす**ように設計されているが，いくつか違った機能を果たすよう組合せることもできる．

For best results, the reaming process has to **be** properly **performed**. Typical application involves bore 25 mm in diameter and 105 mm long in a cast iron part. The reaming **is done** in two passes, the first at a feedrate of 0.6 to 1.0mm/min and the

最も良い結果を得るためには，リーマ加工作業を適切に**行なわねばならない．**鋳鉄部品の直径25mm，長さ105mmの穴はその例である．このリーマ加工は，2回通しで**し**，最初は送り早さ

second at 2.5 to 4.0m/min.

0.6~1.0mm/min で，2回目は2.5~4.0m/min である．

All operations at station 4 **are performed** manually.

ステーション4のすべての作業は，マニアル**操作**である．

Several other machining operations can **be performed** on a drill press.

その他若干の切削作業は，ボール盤で**することができる．**

These operations may **be performed** in a sequence on more than one side of a workpiece without changes in set up.

段取りを変えずに，工作物の1つ以上の側面に，順次これらの加工を**する**ことができる．

Although these air-hydraulic presses are compact in size, they can **perform** operation on metals, plastics, and ceramics that are currently **being performed** on much larger presses.

これらの空気・液圧プレスの大きさはコンパクトであるものの，現在は遙かに大きなプレスで**行なっている**金属，プラスチックおよびセラミックの加工を**する**ことができる．

Progressive drilling operations **are performed** in the center of one end of rods at stations two, three, and four.

2，3，4ステーションで，ロッドの一端の中心に，順次穴明け作業**をする．**

Bending **is** commonly **performed** in one of three ways.

曲げ加工は，普通3つの方法のうちの1つで，**行なわれる．**

Because the analytical work can **be performed** in the field, results are available within an hour after sampling is completed.

現場で，解析作業を**する**ことができるので，サンプリングをしてから1時間以内に結果が手に入る．

A spiral broach has been designed and built for cutting internal helical gears, and turret broaching machines can **perform**

特殊ブローチは，内歯はすば歯車切削用に設計・製作されており，タレット・ブローチ盤は，

multiple operations through automatic indexing.

自動割出しにより，複数の作業を**する**ことができる．

Programming **is performed** through a hand-held control.

プログラミングは，ハンドヘルド・コントローラで**行なう．**

Basic maneuvers of Columbia **are** all **performed** through computers.

コロンビア号の基本的操航は，すべてコンピュータによって**行なわれる．**

Part alignment **was performed** with the use of a laser beam.

部品のセンター合せは，レーザ光線を使って**行なわれた．**

Cutting **is** still **performed** with tools manufactured from high-speed steel.

切削は今でも，高速度鋼でつくったバイトで**行なわれている．**

A single robot can **perform** a dozen or more different spotwelds with greater **precision** than a human.

1つのロボットで，人間よりもより高精度で，12以上の異なったスポット溶接を**する**ことができる．

Electro-optical pitch line runout gage **performs** inspection function without touching the parts being gaged and without using a specialized laser light.

電子・光学的ピッチ線振れゲージは，測定部品に触れることなく，また特殊なレーザ光線を使うことなく，検査機能を**果たす．**

They **are performed** automatically by the System P-Model F.

これらは，システムP-F形で自動的に**行なわれる．**

These may **be performed** by data processing.

これらは，データ処理によって**行なう**こともできる．

Drilling **was performed** to a depth of approximately 10 ft (3m) before inspection and replacement of the drill blade.

ドリルの刃の検査や交換をすることなく，約10フィート（3m）の深さまで穴明け**すること**

ができた．

The computer automatically **performs** the analyses and calculations to determine if the feature is within tolerance.

形状が許容差内にあるかどうかを決定するために，コンピュータが自動的に解析と計算を**する**．

practice : to do something actively, to carry out in action	**する，実行する**

Molten salt bath nitriding **is practiced** world wide by the German process known as "Tenifer" or "Tufftride".

溶融塩浴窒化は，テニファまたはタフトライドとして知られているドイツ式処理法で，世界的に広く**実施されている**．

Put the plan **into practice**.

計画を**実行に移す**．

take : to find out and record, to perform, to deal with	**する，処理する**

If a potential problem exists, the system alerts the operator and identifies the source of the problem so preventive or corrective action can **be taken** immediately.

もし問題が潜んでいれば，ただちに予防あるいは訂正**処置をとる**ことができるよう，オペレータに警告し，かつその問題の原因を明らかにする．

In one case, only two tests **were taken**.

ある場合は，試験を２つ**した**だけである．

Duo **takes** heavyduty <u>cuts</u>.

重<u>切削</u>**する**．

After **taking** one or several roughing cuts (depending on the diameter of the workpiece), 0.015 to 0.030 in. should be left for finishing.

１回または数回荒切削**した**あと（工作物の直径を考慮して），仕上加工用に0.015～0.030インチ残すこと．

Take long, flowing **strokes** with the brush, holding it so that it is inclined at about 30 degrees to the surface being treated and working in the same direction throughout.

処理表面に約30°傾くようにブラシを保持して，終始同じ方向で，長く流れるように**なでる**.

Reading **can be taken** while cutting is in process.

切削中に，読取ることが**できる**.

undertake : to agree or promise to do something

する

The operation should **be undertaken** only if the flywheel is of steel, as cast iron has a tendency to burst.

鋳鉄は破裂する傾向があるので，フライホイールが鋼のものであるときだけ，この作業を**する**.

Reassembling the gearbox will **be undertaken** in the reverse order to dismantling.

歯車箱の再組立は，取外しの逆の順序で**する**ことができるであろう.

確かめる（確認）

analyse : to examine and interpret

調べる（解明，解析，分析）

Aanalyse the causes of their failure.

その失敗の原因を**調べる**.

The results **were analysed** statistically, and plotted against every possible characteristic of the grease.

結果を統計的に**解析**し，グリースの考えられるあらゆる特性についてプロットした.

A detailed **analysis is done**.

詳細に**解析する**.

The wear volume measurements exhibit significant scatter and a statistical signifi-

摩耗量の測定値は著しくバラツキがあるので，データを統計

cance **analysis** of the data **was made** .

的に**解析**した．

From **analyses** on liquid remaining on substrate, ...

母材に残った液の**分析**により…．

---

check : to test or examine in order to make sure that something is correct or in good condition.

調べる，確かめる（点検，検査）

This new automatic tester automatically **checks** the hardness of parts and shows the results on a digital read out, a recorder, indicator lights by separating the good parts from the bad.

この新しい自動テスターは，部品の硬さを自動的に**チェック**して，良い部品を悪いものから分けて，その結果をディジタル読取り器，記録計，指示灯に示す．

Check that replacement bearings are readily available in case they are needed.

代わりの軸受が，必要な時にすぐ入手できることを**確かめる**．

While the plug is being examined the spark gap should **be checked**.

プラグを調べている間に，スパークギャップを**確かめる**こと．

If there is a slightest doubt as to bearing performance, the machine should **be** stopped and the bearing arrangement **checked**.

軸受性能に少しでも疑わしい点があったら，機械を止めて，軸受の取付け状態を**チェックする**こと．

The wiring and connections should **be** carefully **checked over**.

配線および接続を，注意深く**くまなく確かめる**こと．

It is always best to **check up** the faces on a surface plate before fitting a thin gasket, or there will be a risk of a "blow" occurring.

薄いガスケットを取付ける前に，表面板の面を**調べ上げる**ことが常に最善で，そうしないと「吹き抜け」が起こる惧れがある．

Turn boost switch OFF to functionally **check** solenoid valve.

ソレノイド弁の機能を**確かめる**ために，ブーストスイッチを切る．

Parts that give evidence of wear or physical damage will **be checked** dimensionally.

摩耗や物理的損傷の跡のある部品は，寸法を**チェックする**．

Visually **check** the condition of a cable.

ケーブルの状態を目視**検査する**．

Check for loose electrical connections.

電気的接続部の緩みを**確かめる**．

Systems should **be checked** daily for correct air and oil temperature, oil level and mist header pressure.

システムのエア，油の温度，油面およびミストヘッダ圧力が正しいかどうか，毎日**チェックすること**．

Check if the sensor is working properly.

センサが適切に働いているかどうか**確かめる**．

Check the condition of the seals near the bearings to ensure that they will not, for example, permit hot or corrosive liquids and gases to penetrate the bearing arrangement.

軸受に近いシールの状態，たとえば，シールが熱い液あるいは腐蝕性の液を間違いなく軸受装置に浸透させていないことを**確かめる**．

In the control area of the system, the timer should **be checked** to assure that the predetermined time settings have not been altered.

システムのこの制御部分では，あらかじめ決めた時間の設定が変わっていないことを確かめるために，タイマを**チェックすること**．

Any automatic lubricating devices should **be checked** to see that they function

どんな自動給油装置でも，それが正しく機能していることを

correctly.

見て**確かめること．**

To **check** the accuracy of the work done in enigeering the system, pick a bearing at random in Fig. 5, then **check** it against the discharge capacity and adjustment setting of the dual line measuring valve selected.

システムの設計・施工で行なった作業精度を**チェックする**ために，図５のように，無作為に１つの軸受を抽出して，最良の計量弁のセッティングの調整および排出容量を**チェックする．**

Chamfers **are checked** by means of special templates.

面取りは，特殊なテンプレートで**チェックする．**

Check the temperature of the bearing arrangement by using a thermometer, heat sensitive chalk, or often simply placing a hand on the bearing housing.

この軸受装置の温度は，温度計，感熱チョークを使うか，あるいはよく行なわれているように軸受ハウジングに単に手を当てて，**確かめる．**

Check the inside of the spindle taper with your finger for nick or grit.

スピンドル・テーパの内側を，傷や砂粒があるかどうか手で**チェックする．**

The height of single-acting thrust bearings **can be checked** between a surface plate and the gauging pin, or with a micrometer.

単列スラスト軸受の高さは，定盤と測定ピンの間またはマイクロメータで**チェックできる．**

Where a non-standard bearing is to be specified, the availability should **be checked** with S.

標準でない軸受を指定するような場合には，入手の難易をＳ社に**確かめること．**

Perform operational **check**, observe for leaks.

作動**チェックをし**，漏れを調べる．

Checking the various points at which play occur **is** most easily **carried out** with

遊びを生じているいろいろな点の**チェックは**，１人の助手に

the aid of an assistant.	手伝ってもらえば，最も簡単に**できる．**
While the pumps run, **make** the hourly **check** listed below : bearing temperature, suction and discharge pressure.	ポンプ運転中は，下記を1時間ごとに**チェックする**：軸受温度，吸入および吐出圧力．
A final **check** should **be made** to be sure all clamping bolts are tight.	固定ボルト全部が間違いなく，固く締まっていることを最終**チェックすること．**
A conventional **check** on maximum speed limits **can be made** from a dn value.	限界最高速度の**チェック**は普通 dn 値で**できる．**
Keep a check on bearing behavior immediately after starting up.	始動直後は，軸受の挙動をずっと**チェックしていること．**
……**run** 10,000 electrical **checkes** a second.	……は毎秒10,000回の電気的**チェックをする．**

〈用　語　例〉

automatic check　自動検査
check bit　検査ビット，チェック・ビット
check of drawing　検図
marginal check　限界検査
programmed checking　プログラムによる検査
redundancy check　冗長検査
thread plug gage for checking wear　摩耗点検ネジプラグ・ゲージ
total check　全体検査

validate : to make valid, to confirm	**確かめる（確認）**

Such methods must **be validated** by friction measurements.	このような方法は，摩擦測定で**確かめねばならない．**

verify : to check the truth or correctness of	**調べる，確かめる**

M has three models in its Series 9, the	M社には，シリーズ9に3形

smallest accepts workpieces 14″ long and 7″ in diameter; the largest **verifies** workpieces 40″ long and 16.4″ in diameter.

式あり，最も小さなものは工作物の長さ14インチ，直径7インチ，最も大きいものは長さ40インチ，直径16.4インチの工作物を**検査**できる．

Refractometer accuracy should also **be** periodically **verified**.

屈折計の精度は定期的に**確かめる**こと．

The value **is** independently **verified** before the specimen is tested.

試料を試験する前に，この値を個々に**確かめる**．

Before attempting to remove the pinions they should be carefully inspected for wear and backlash, and it should **be verified** that timing marks are provided on the teeth.

ピニオンを取外そうとする前に，摩耗とバックラッシュを検査する．そしてタイミング・マークが歯についていることを**確認する**こと．

The readout accessory **verifies** position for X and Y axes to assure precise hole location.

この付属読取り装置は，精密な穴位置を確保するために，X軸およびY軸の位置を**確認する**．

The operator then removes an appropriate shim from the rack and **verifies** its size before adding it to the differential assembly. **Verification is accomplished** by inserting the shim into an automatic gage.

つぎに，作業者は架から適当なシムを取出して，デフ組立品に追加する前にその寸法を**確かめる**．**確認**はシムを自動ゲージに挿し込むことによって**行なう**．

Calibration change is not required for different part weights. A spindle lock feature provides on-machine correction and instantaneous **verification**.

部品重量の違うものについても，補正を変更する必要はない．スピンドル固定装置は，機械に取付けたままで修正および即時に**確認**できる特徴がある．

〈用　語　例〉

verification tolerance　検定公差　　verifier　検孔機

たてる

tapping : tap ; a device for cutting a screw-thread inside a cavity.　タップ立て，ネジ立て，切り付け

Tap holes by hand or with a drill press.

手（で）あるいはボール盤で，穴に**タップを立てる**．

Unlimited continous **tapping** operations can be performed without subjecting hydraulic motor to any undue strain.

油圧モータを過度に歪ませるようなことなく，無制限に連続**タップ立て**作業が行なえる．

Standard plug or bottoming taps can be used when hand **tapping** in the lathe. If power is used, what kind of tap works best ?

旋盤において，手で**タップ立てする**場合は，標準プラグや仕上タップを使うことができる．もし，動力を使うとしたら，どんな種類のタップが最も良いか？

〈用　語　例〉

automatic nut tapping machine　ナットネジ立て自動盤，ナットタッピング機

centering machine　心立て盤，心取機

tap borer, tap drill　ネジ下切り

tap-end stud〔bolt〕　植込みボルト

tapping machine, tapper, threading machine　ネジ立て盤

tapping screw　タッピングネジ

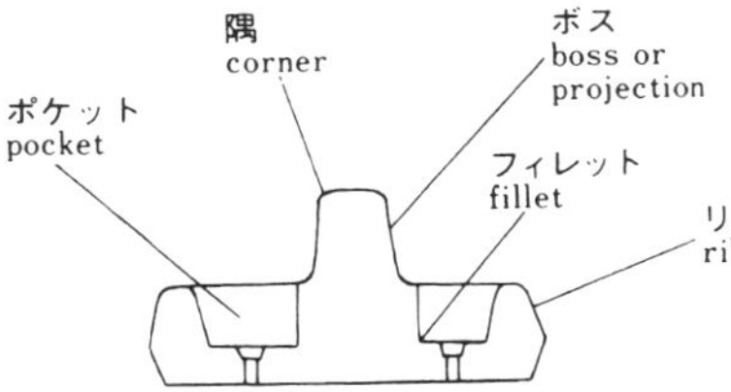

調整する，(〜を)合わせる，調節する，整える

adjust: to arrange to put into the proper position, to alter by a small amount so as to fit or be right for use	調整する，加減して合わせる

Adjust the zero point without anything on the balance pan.

天秤皿に何も載せないで，ゼロ点を**調整する**.

It may be necessary to **adjust** the spacer more than once.

スペーサを一度ならず，**加減して合わせる**ことが必要であろう.

Data is completely consistent and time-corrected. You can continuously **adjust** operations in terms of efficiency, cost, safety and health.

データは完全に一定し，時間修正もされているので，作業を効率，コスト，安全，健康に関して絶えず**適正な状態に合わせる**ことができる.

Adjust the constant temperature bath to 10℃.

恒温槽を10℃に**調整する**.

Adjust the inside and outside mercury height to same level watching the buret.

ビュレットを見ながら，内側と外側の水銀柱を同じ高さに**合わせる**.

Tooling can **be adjusted** to a general length and diameter range, and then the axes can **be** fine **adjusted** to the tool point.

ツーリングは全般的な長さおよび直径範囲内に**調整**でき，ついで軸線をバイト先端に微**調整する**ことができる.

Taper roller bearings may be fairly complicated to mount. Often they have to

円錐コロ軸受は，取付けるのにかなり面倒なことがある．こ

be adjusted to a certain amount of internal clearance or to a given preload using springs or shims (calibrated distance piece).

の軸受は，一定量の内部隙間にあるいはバネかシム（補正済間隔筒）を使って所定の予圧になるように，しばしば**調整**しなければならない．

The length of stroke can **be adjusted** to suit the workpiece.

ストロークの長さは工作物に合うように**調整**できる．

The lubricator can **be adjusted** to meet the requirement of your tool.

潤滑器は，ツールの要求条件に合うように**調整**できる．

Adjust the eye piece so that the cross hairs is seen clearly and easily.

十字線がはっきり，そして楽に見えるように，接眼レンズを**調節する**．

Loosen screw and **adjust** the regulator plate so that the latch lever clears the pawl lever by 1／16″.

ネジを弛めて，ラッチ・レバーが爪レバーから1／16インチだけ離れるように，調節板を**調整する**．

Adjust the indicator so that the pointer indicates zero.

針がゼロを指すように，インジケータを**調整する**．

Indicator **is adjusted** so that it registers zero.

記録計がゼロを表示するようにインジケータを**調整する**．

Adjust check nut for 0.005 to 0.015 inch clearance with washer.

座金で，隙間が0.005～0.015インチになるように，チェック・ナットを**調整する**．

Adjust clevis on piston of cylinder assembly for correct length and lock with nut.

シリンダ組立品のピストンのクレビスを正しい長さに**調整**し，ナットで固定する．

The clamp automatically **adjusts** for

クランプは，加工物の高さ変

variations in workpiece height.

化に対し，自動的に**適合する**.

Extending from the head is a 2″ non-rotating, spindle which can **be adjusted** for length, thus giving the micrometer an overall length of $7\frac{1}{8}''$ ～ $9\frac{1}{4}''$.

ヘッドから突出しているのは，長さを**調整**できる2インチの非回転スピンドルで，これでマイクロメータは全長$7\frac{1}{8}$～$9\frac{1}{4}$インチになる.

The conveyor's variable-speed drive can **be adjusted** from 0～20fpm.

コンベヤの可変速駆動装置は，0から20fpmまで**加減**できる.

Single row angular contact bearings **are adjusted** towards another bearing.

単列アンギュラ軸受は，もう1つの軸受の方向に**調整する**.

Clearance **is adjusted** by means of the spindle nut unit until the torque is felt to increase.

隙間を，トルクの増加が感じられるまで，スピンドル・ナットユニットで**調整する**.

Both axes can **be adjusted** by hand-wheels.

両軸とも，ハンドホイールで**調整**できる.

The oil pressure can **be adjusted** by

油圧は，**調整**ネジを回わして

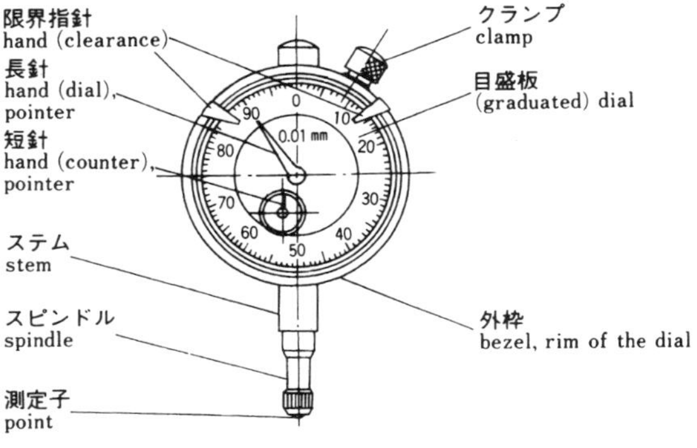

0.01㎜ 目盛ダイヤルゲージ
(dial gauge reading in 0.01 ㎜) (dial indicator)

turning the **adjusting** screw in to increase or out to decrease pressure.

増圧したり，減圧したりの**調節**ができる．

If desired, the numerical value of A can **be adjusted** up and down by varying the parameter M and Z.

希望により，パラメータMおよびZを変えることによって，Aの数値を上げたり下げたり**加減**できる．

Each jaw can **be adjusted** independently by rotation of the radially mounted threaded screw.

それぞれのジョーは，半径方向に取付けたネジを回わすことによって，別々に**調整**できる．

The slide can also **be** manually **adjusted** during rotation by means of a hand crank. The **adjustable** vernier enables the operator to **adjust** the tool bit with the hand crank by 0.00025″ per line (0.0005″ on diameter).

スライドは，ハンドクランクで回転中に手でも**調整**できる．**調整**バーニヤにより，作業者は1目盛について0.00025インチ（直径で0.0005インチ）ずつ，ハンドクランクでツールを**調節**できる．

The cutaway drawing shown here illustrates the mechanical／electrical components inside the turret and slide gearbox. There are no limit switches to set, valves to **adjust** or knobs to turn.

ここに示す切断図は，タレットおよび送り台歯車箱の中の機械的・電気的部品を図示したものである．セッティング用リミットスイッチ，**調整**弁，回転ツマミは，まったくない．

Don't attempt to **make** the **adjustment** while the pump operates.

ポンプが動いているときは，**調整し**ようとしないこと．

0.606 feeler gauge can be used in **making** this **adjustment**.

この**調整をする**のに，0.606のすきみゲージ計が使用できる．

Operating with either jaw or collet chucks, variable tension **adjustment** of 0.030″ **is made** by hand, even while in

ジョーまたはコレットチャックでの作業は，作業中でも張力を0.03インチ手で**調整**できる．

operation.

The **adjustment is made** manually and simply by means of a screw located at the top of the columns.

この**調整は**，コラムの頂部にあるネジによって，人手で簡単**にできる**.

Adjustments on the jaw can **be made** with a 1／4″ hex wrench.

ジョーの**調節**は，1/4インチ六角レンチで**できる**.

It is necessary to **effect re-adjustment**.

再調整をすることが必要である.

After **adjustments** are complete and engine shut down, accomplish the following:

調整を済ませ，エンジンを停止してから，つぎのことをする：

The generator attaching brackets are slotted to **permit** this **adjustment**.

この発電機取付けブラケットは，この**調整ができる**ように溝が切ってある.

Suitable **adjustment is obtained** by means of shaft nuts shimming.

軸ナットにシムを入れることによって，適切に**調整できる**.

Employ different **adjustment**.

違った**調整**法でする.

A calculation can be made of how much **adjustment** the bearings still need to give the required clearance.

所望の隙間にするのに，軸受の**調整**がまだどれだけ必要か計算できる.

All 22 bearings in the experiments operated satisfactorily for the duration of the tests although some minor **adjustments** were required in the radial temperature gradients of the bearings in test Number 11 to avoid high running torques.

高運転トルクを避けるために，試験番号11の軸受の半径方向の温度勾配について若干の**調整**が必要であったが，実験の軸受22個全部がこの試験時間の間申し分なく運転できた.

Screw adjustment increases versatility and permits use of cams from older design machines.

ネジ調整によって汎用性が増え，旧式機械のカムを使用することもできる．

Head adjustment up to 45° front and back and 30° left and right permits rapid milling from any angle.

前後45°，左右30°までの**ヘッド調整**で，どんな角度からでも迅速なフライス加工ができる．

Turn each pressure **adjustment screw** counterclockwise until restricted.

圧力**調節ネジ**１つ１つを，回わせなくなるまで反時計方向に回わす．

Adjustment screw with dial graduated to 0.001 must be screwed into the assembly as far as possible.

0.001に目盛ったダイヤル付**調整ネジ**を，アッシにできるだけ深くねじ込む．

Three vertical spindle **adjustable** in height.

高さ**調整のできる**３つの立形スピンドル．

……**adjustable** to any height.

……どんな高さにも**調節できる**．

The spindle is carried in a head that is vertically **adjustable** on the column, being provided with down feed by means of worm gearing.

このスピンドルは，コラム上を垂直方向に**調節できる**ヘッドの中に取付けられていて，ウォーム歯車によって下方送りが与えられる．

Infinitely **adjustable** between 100 and 640rpm.

100～640rpmの間，無段**調節**．

Back-up plate is **adjustable** for control of finished part thickness.

バックアップ・プレートは，仕上げ部品の厚さを制御できるように**調整**できる．

Adjustable holders for the cross slide

クロススライド・カムの**可変**

cams permit changing their positions in relation to each other or to the lead cam.

ホルダは，相互あるいは親カムとの位置関係を変えることができる．

The complete system includes: 0 to 5,000 pressure gage, oil dip stick, **adjustable** air pressure regulator,…….

全システムには，0～5.000圧力計，検油棒，**可変**空気圧調節器……などがある．

The crank-**adjustable** furniture.

クランク**調整**の調度品．

The servo control system features a servo spindle speed **adjustor**.

サーボ制御システムには，サーボ・スピンドル速度**調整器**が付いている．

〈用　語　例〉

adjust screw　調整ネジ
adjustable bed crank press　可動ベッド形クランクプレス
adjustable curve rule　自在曲線定規
adjustable hand reamer　可調整ハンドリーマ
adjustable spanner　自在スパナ
adjustable restrictor valve　可調整制止弁
〔adjustable round〕 split die, button die　調整丸ダイス
adjustable bearing　調整軸受
adjusting bolt　調整ボルト
adjustable reamer　調整リーマ
adjusting screw　調節ネジ，調整ネジ
adjustment "zero" feeder　送りのゼロ点調整
sampling inspection with adjustment　調整式抜取り検査

control: to regulate, to restrain, to manage, to direct

制御する，管理する，調節する

Adjustment screw **controls** lubricant output within the limits of valve size.

調整ネジは，弁寸法の限度以内で潤滑剤の排出量を**調節する**．

The units possess the following operational capabilities: multiple spindle heads with the capacity for 2, 3 or 4 drill heads with a choice of fixed or adjustable-type

このユニットはつぎの作業が可能である．すなわち，固定式または調整式チャック付きの2，3または4ドリル・ヘッドが付

chuck, a multi-Stepfeeder **control** to regulate feed and recoil motion for drilling holes of depths to five and six times greater than the drill bit diameter, ……, and a Hydro-Speed regulator to **control** the forward speed at the most suitable rate for the workpiece material.

けられる多軸ヘッド，ドリル・ビット直径より5，6倍までの深さの穴明けができるよう，送りおよび巻戻し運動を調節するためのマルチ・ステップフィーダ**制御**，……，およびワーク材料に最も適した速度に前進速度を**制御する**ハイドロスピード調節器．

Newcomer molded chipbreaker inserts **control** the turnings from the workpiece by breaking continuous chips into short lengths.

新しいインサート形モールド・チップブレーカは，連続する切屑を短かい長さに破断することによって，ワークから出る切削屑を**コントロール**する．

The point at which the upset occurs is adjusted according to the total thickness of the two material being jointed. Also, the width of the rivet head is determined by the adjustments that **control** the amount of force generated between the punch and anvil.

据込みが発生する点は，結合されている2つの材料の全厚さによって加減される．またリベット頭の幅は，パンチとアンビル間に生じる力を**制御する**調整によって決まる．

Feed speeds can **be controlled** independent of spindle speed.

送り速さは，スピンドル速度と無関係に**コントロール**できる．

Controlling techniques available include pendant pushbutton operator **controlled**; fixed sequence automatic operation by series of electromechanical switches and relay logics; and…….

利用できる**制御**技法には，オペレータ**操作**の吊下げ式押ボタン，一連の電子・機械的スイッチおよびリレー・ロジックによる固定シーケンス自動制御，……などがある．

A multitude of methods are used to **control** wear. But each in itself is very specific and has limited applicability to the

摩耗を**コントロールする**ために使われている方法は数多い．しかし，それぞれがきわめて特

particular situation for which it was devised.

定のもので，適応性はそれ用に考案された特殊状況に限られる．

The leadscrew **controls** sidewise movement of the saddle which is attached to the table.

親ネジは，テーブルに取付けられたサドルの横方向の動きを**コントロールする．**

The shear stresses at the surfaces and their dependance on lubricant EHL film thickness **control** crack initiation and growth.

表面における剪断応力，および潤滑剤 EHL 膜厚によって，亀裂の発生および成長が**左右される．**

The furnace temperature **is** automatically **controlled** with a temperature control instrument.

炉の温度は，温度制御器で自動**制御される．**

Bearing operating temperature can **be controlled** by adjusting oil flow rate.

軸受運転温度は，油の流速を加減することで**制御**できる．

Spindle speed **is controlled by** a solid state, variable frequency AC drive.

主軸速度は，ソリッド・ステート可変周波数 AC 駆動装置で**制御する．**

The carriage is mechanically locked in position. Control signals and position feedback **are controlled** by the robot's microcomputer.

往復台は所定位置に機械的に固定される．制御信号および位置フィードバックは，ロボットのマイクロコンピュータで**制御される．**

A clear scale enables point angle to be set, and clearance **is controlled** by a lever behind the chuck.

明瞭な目盛で先端角を合わせることができ，逃げはチャックの後のレバーで**加減する．**

The speeds and feeds **are controlled** by the width of the tool in relation to the work diameter, the amount of overhang,

速度および送りは，ワークの直径，オーバーハング量および輪郭の形状を考慮した工具幅に

and the shape of the contour.

よって**制約される**.

Reamer cutting action **is controlled** to a large extent by the cutting speed and feed used.

リーマの切削作用は，使用する切削速度および送りによって，大幅に**制約される**.

All machine functions can **be** tape-**controlled** for automatic operation and manually **controlled** by dials.

機械のすべての機能は，自動操作ではテープ・**コントロール**でき，手動ではダイヤルで**コントロール**できる.

These new welders **are** computer **controlled**: they are fully automatic robots.

これらの新しい溶接機は，コンピュータ**制御**である．すなわち，完全自動ロボットである.

The diffusion process **is** rate-**controlled**.

拡散処理は速度**制御**である.

The mechanical indexing rotary table **is** cam **controlled**.

機械的割出し回転テーブルはカム**制御**である.

Hydraulic chuck **controlled** from apron or head stock…….

エプロンまたは主軸台で**制御する**油圧チャックは…….

The radial thermal gradients **were** very closely **controlled** to regulate the ball-to-raceway contact stresses of the duplex bearing pairs.

対の組合せ軸受の玉と軌道の接触応力を規制するために，半径方向の温度傾斜を，非常に厳密に**管理した**.

The rise and fall of the bed **is controlled** by a pushbutton, for raising and lowering a total of 54 inch at 10 ipm.

ベッドの昇降は，速度10ipmで全体長54インチの上げ下げができるよう押ボタンで**制御する**.

〈用　語　例〉

controlled atmosphere heat treatment　雰囲気調節熱処理

controlled initial unbalance　調整済み初期不釣合い

control wheel　調整車（調整砥石，送り砥石）

error control system　誤り制御方式

control level　管理水準
control panel　制御盤，操作盤
control rod　制御棒，引棒，調節棒
oil level controller　油面調整装置
volume control pressure pump　容量式圧力制御ポンプ

dial: to select or regulate by means of a dial　**ダイヤルで調節する**

The diameter of the workpiece is checked with a micrometer and the remaining amount to be cut **is dialed** on the cross feed micrometer.　加工物の直径をマイクロメータでチェックし，切削すべき残量を横送りマイクロメータに**ダイヤルで設定する．**

focus: to adjust the focus of (a lens or the eye)　**焦点を合わせる**

Cover the head with a black piece of cloth to **focus**.　**ピントを合わせる**のに黒い布片で頭を覆う．

Focus the telescope to infinite distance.　望遠鏡の**焦点を**無限大にする．

For measurement, **focus** on a mark on the object.　測定には，物体のマークに**焦点を合わせる．**

regulate: to adjust or control (a thing) so that it works correctly or according to one's requirements
: to control or direct by means of rules and restrictions　**調整する，調節する**

A **regulating** roll drives the work and **regulates** its speed.　**調整**ロールはワークを駆動し，かつその速度を**調節する．**

Concentration and lubricity of the solution will **be** measured and automatically **regulated**.　溶液の濃度および潤滑性は，測定されかつ自動**調整**される．

Feed speed of saw can **be regulated**　鋸の送り速度は 0 と590ipm

between 0 and 590 ipm and is adjusted from front mounted controls.

の間で**調節**でき，前面に取付けた制御装置で加減する．

It is not always possible to **regulate** the amount of oil to the extremely small quantity desired.

油の量は，所望のごく少量にいつも**調節**できるとは限らない．

Regulate the temperature so as to be at about 5～8℃.

温度を，約５～８℃であるように**調節する**．

The amount of heated air **is regulated** by a valve mounted through the firewall.

加熱空気量は，防火壁を通して取付けた弁で**調節する**．

Tight speed **regulation is obtained** by tachometer feedback.

厳格な**速度制御**は，タコメータのフィードバックによって**できる**．

Centralized control can range from a simple on-off control of one device to the highly complex **regulation** of a complete building air conditioning system.

集中制御には，１つの装置の簡単なオン・オフ制御から，ビルの空調システムのように非常に複雑な**調節**までである．

〈用　語　例〉

feed regulator friction washer　送り調節座金
flow regulating valve　流量調整弁
pressure regulating thumb screw　押さえ調節ネジ
regulating wheel truing device　調整車形直し（ツルーイング）装置

set : to adjust the hands of (a clock) or the mechanism of (a trap etc.), to put into a specified state.　**合わせる，セットする**

Set the dial reading to…….

……にダイヤルの読みを**合わせる**．

Magnification of errors can **be** dial **set** to 100×/500×/200× or 1,000×.

誤差の倍率は，ダイヤルで，100倍，500倍，200倍または1,000

倍にセットできる.

The tool may **be set** to the dead center or to a steel rule on the workpiece.

バイトは，止まりセンタまたは加工物に当てた鋼製定規に**合わせる**ことができる.

The engine **is set** to run at the speed.

エンジンをこの速度で運転するように，**セットする**.

tune: to adjust (an engine) to run smoothly
調子を整える

With this up-to-the-minute information, you can **tune** for much lower-cost operation——minimum energy consumption and the best material balance.

このほやほやの情報で，遙かに安いコストの操業——最低のエネルギ消費と最良の材料バランス——ができるように**調整することができる**.

Today, our company has a finely **tuned** computerized system of analysis of its lubricant consumption and costs.

現在，わが社は，潤滑剤消費量およびコストの解析を詳細に**調整した**コンピュータ化システムをもっている.

ついや(費)す

consume: to use up
使い尽す，消費する

……**consume** very little electricity.

……は，電力**消費**がきわめてわずかである..

When the belt is fully loaded, the motor **consumes** 3.8 percent less power with the synthetic oil in service.

ベルトが全負荷のとき，この合成油を使うと，モータの動力**消費**は3.8%少ない.

Electrical power **consumed** by the

切削作業中に工作機械が**消費**

machine tool during a machining operation is …….

する電力は…….

The use of T-fittings for branching tube or pipe is also costly and **time consuming**.

チューブやパイプを分岐するのにT形金具を使うことは，コストと**時間がかかる**.

expend : to spend (money, time, care, etc.), to use up

ついやす

Of this friction energy, about 60～80 percent **is expended** at the piston／cylinder wall interface.

この摩擦エネルギのうち約60～80%は，ピストンとシリンダの接触面で**ついやされる**.

spend : to use for a certain purpose, to use up

使う，ついやす

Time **spent** by the operator on loading the workpiece…….

工作物をローディングするのにオペレータが**ついやす**時間は…….

A considerable amount of time and effort must **be spent**.

かなりの量の時間と努力を**使わ**なければならない.

take : to consume

使い尽す（消費する）

A small portion of the energy output **is taken up** by friction.

この出力エネルギのうちのわずかな部分が，摩擦によって**ついやされる**.

use up : to use the whole of (material, etc.)

使い尽す

Wood in England began to **be used up** …….

イングランドの木は，**使いきられ**始めた…….

waste : to use extravagantly or needlessly or without an adequate result — **むだに使う（消耗する）**

……**is wasted** gradually by radiation.

……は，放熱によって次第に**消失する**.

All sources of energy upon which industry depends **are wasted** when they are employed ; and industry is expending them at a continually increasing rate.

工業が依存するすべてのエネルギ源は，それが使われるときに**消耗される**；そして工業は絶えずその割合いを増やしながらエネルギ源を消費し続けている.

The thermograph shows the **waste** of energy in making of power.

サーモグラフは，動力をつくる際のエネルギの**無駄**を示している.

This accuracy and plus complete reproducivility can eliminate the **wastage** caused by human error.

この精度と完全な再現性とによって，これまでの人的エラーによる**無駄**をなくすことができる.

つかう，使う，用いる

adopt : to take and use as one's own — **使う，採用する**

Shaping by rolling **is adopted**.

ロールによる成形法を**採用する**.

To remedy this defect, automobile-type gears of hardened nickel chrome steel **were adopted**.

この欠陥を直すのに，焼入れ NiCr 鋼の自動車用歯車を**採用**した.

Sellers proposed a range of screw threads and bolt sizes which **were adopted** as standard by the U.S. government in

セラーは，ある範囲のネジとボルトの寸法を提案した．そしてこれを1868年米国政府が，規

1868.

格として**採用**した.

The method **adopted** limits the form of software aids that can **be used**.

使用方法によって，**利用**のできるソフトウェア手段の様式が限定される.

apply : to bring into use — **使う，適用する，応用する**

This manufacturer **applied** tolelance three to four times greater than available.

このメーカーは，普通使うものより 3 ～ 4 倍大きい公差を**適用**した.

By **applying** the method developed by M, ……

Mが開発した方法を**使う**ことによって，……

A properly **applied** and installed oil mist system can reduce maintenance costs significantly by reducing bearing failures as much as 90 percent.

使い方および取付け方法が適正なオイルミスト・システムによって軸受の損耗を90%減らすことにより，保全コストを顕著に低減できる.

The planer is the machine tool **employing** single-point cutting tools, and some machines are designed to **apply** multipoint cutting tools to the workpiece.

平削り盤は，1 刃のバイトを**使う**工作機械である. 工作物に対し多刃の工具を**適用する**ように設計した機械もある.

A system of auto-control **is applied** to mill.

自動制御システムをフライス加工に**使う**.

The process for reducing the thickness of aluminum sheet in hot or cold rolling mills requires that a fluid be **applied** to the strip and rolls to act both as a lubricant and a coolant.

熱間または冷間圧延機でアルミの薄板の厚さを薄くする工程では，潤滑剤および冷却剤の両方として作用するように帯板とロールに，液を**かける**ことが必要である.

……, which is of considerable use, **is being applied** to …….

……は，かなり有用で……に**使われている**．

The first method for the measurements of wear **was** generally **applied** to the experiments where there was very small removal and the second **was applied** for the relatively greater wear conditions.

摩耗測定の第一の方法は，摩耗量がきわめて小さい場合の実験に**使い**，第二の方法は比較的大きい摩耗条件の場合に**使った**．

The Fastitch process is being successfully **applied** on a wide variety of parts in such fields as automotive, office equipment, appliances, and electrical equipment. In the automotive field, **it's used** on weatherstripping and various moldings, and on the flange of aluminum-steel hood panels.

このFastitchプロセスを自動車，事務用機器および電気装置などの分野の広範な部品に**使って**成果をおさめている．自動車の分野では，ワイパおよび各種モールド品，およびアルミ-鋼フードパネルのフランジに**使われている**．

Cooling water system **is applied** in diesel engine.

水冷方式は，ヂーゼルエンジンに**使われている**．

Felt ring **should** where possible **be applied** as continuous ring.

フェルト環は，できれば，連続リングで**適用すること**．

One device is the spring washer commonly **applied** as shown in Fig. 2 .

1つの道具は，図2に示すように普通**使われる**バネ座金である．

…… will **receive** more general **application**.

……は，もっと一般的に**使われる**であろう．

These methods **find** wide **application** in practice.

これらの方法は，実際に広い**用途がある**．

They **find** widest **applications** in machining of cast iron, malleable iron, and

鋳鉄，可鍛鉄および硬質青銅の切削に，広い**使いみちがある**．

the hard bronze.

The **application** of a lubricant to a machining operation often reduces the tendency for a built-up edge to form.

潤滑剤を切削作業に**使う**と，構成刃先の形成を減らすことが多い.

Additives such as sulfur, chlorine, and other chemicals can be used to improve lubricating and EP properties, but these fluids have only limited **application** for grinding these materials.

硫黄，塩素その他の薬品のような添加剤は，潤滑および極圧性を良くするのに使うことができるが，これらの液の**用途**は，これら材料の研削だけに限られている.

employ: to make use of (instrument, means, material), to use	**使う**

Grinder **employs** an abrasive wheel.

研削盤は，砥石を**使っている.**

Employ standard machined features whenever possible.

可能なときには，標準の機械加工した形状のものを**使う**こと.

Turret lathes **are** widely **employed**.

タレット旋盤は広く**使われている.**

Feeds of 0.006″ to 0.012″ per revolution **is employed**.

1回転当り0.006～0.012インチの送りを**使う.**

The most popular method **in use** presently **employs** a hydraulic servo valve which contacts a reference point on the workpiece as the tool advances.

現在，**使われている**最もポピュラな方法は，工具が前進すると工作物の基準点に接触する油圧サーボバルブを**使っている.**

The technique **employed** by these machines can save 80 percent boring operation.

これらの機械が**使っている**技法によって，中ぐり作業の80％を節約できる.

GC015 carbide-coated inserts **are employed** on the drill. They withstand high temperatures at the cutting edge and permit fast surface speeds with no fractures.	G超硬被覆のインサートは，ドリルに**使われている**．これは，切刃部での高温に耐えて，高表面速度が可能である．
… can **be employed** on many machine tools.	…は，多くの工作機械に**使う**ことができる．
Conventional transfer lines **employ** single-point turning tools to machine rotor faces.	普通のトランスファ・ラインは，ロータ面の切削に1切れ刃バイトを**使っている**．
Creep-feed grinding, a recent development **used** more extensively in Europe than in the U.S., **is employed** at General Electric's Evendale plant to grind the dovetail serrations in jet engine turbine blades.	GE社のE工場では，最近開発されて，米国よりも欧州で広く**使われている**クリープ送り研削を，ジェットエンジン・タービン翼のダブテール・セレーションの研削に，**使っている**．
The first machine to **employ** plunge milling for brake rotor facing was built by L Machine Tool, Inc.	ブレーキ・ロータの面削りのためのプランジ・フライス削りに**使う**最初の機械は，L社が作った．
The Down-Grip vise **employs** a new principle for holding a workpiece down on the vise bed.	このDown-Gripバイスは，工作物をバイス・ベッドに保持するために新しい原理を**採用している**．
Small lathes **is employed** extensively for work on material in bar form.	小形旋盤は，棒状の材料の加工に広く**使われる**．
Broaching **is** mainly **employed** for machining out holes or other internal surfaces, but can also **be used** for external	ブローチ加工は主として穴など内側表面を機械加工するのに**使われる**が，外側表面および既

surfaces and for burnishing already-formed holes.

製の穴をバニシングすることにも**使う**ことができる．

The same experimental set up **used** in the cast iron research **was employed** for aluminum broaching tests.

鋳鉄品の研究に**使った**同じ実験装置を，アルミニウムのブローチ加工試験に**使った**．

For particular purposes a multi-thread **may be employed** …… e.g., a double or triple threads.

特殊な目的に，たとえば，2条あるいは3条ネジなどの多条ネジが**使われる**．

Welding can **be** usefully **employed** for such unexpected purposes as increasing the diameter of a shaft which has been accidentally turned too small in a lathe or for building up a damaged key-way in a hardened steel shaft.

溶接は，旋盤で小さく削りすぎてしまった軸の直径を大きくするというような意外な目的，あるいは焼入鋼の軸の損傷したキー溝の肉盛をするためにうまく**使う**ことができる．

Under these circumstances, the rolling solution can **be employed** at a lower temperature, thus improving its effectiveness as a coolant.

このような環境下では，圧延用水溶液をより低い温度で**使う**ことができ，冷却剤としての効果を上げられる．

Although EDM had not been **employed** at the Lowell facility previously, other Raytheon plants **had used** it successfully for similar operations,

EDMは以前はL工場では**使われて**いなかったが，他のR工場ではそれを似たような作業に**使って**，うまくいっていた．

Such seals **are** most commonly **employed** where bearing housings are exposed to liquids such as cutting compounds which …….

このようなシールは，切削剤のような液にさらされる軸受ハウジングで，最も一般的に**使われる**．

The law **is employed** as the basis of many recent study of economics.

この法則は，経済学の数多い最近の研究の基礎として**使われている**．

The effect is determined **employing** a laboratory test fixture.	研究所の試験用取付け具を**使って**効果を確かめる．
make the most of : to use the best advantage	**活用する**
… enable programmers to **make the most of** their skills.	…によって，プログラマはその技能を**活用**できる．
Cutting materials will be selected, and developed when necessary, to **make the most of** advanced groove geometries.	この優れた溝形状を**活用する**ために，切削材料を選び，必要なときには新しいものを開発する．
service : use, assistance, a helpful or beneficial act ; be of service, to be useful, to help	**使う，役立てる，役に立つ，扱う**
Designed for those applications where volume of dust and chips does not require frequent removal, the S cyclone-type collector will **service** from eight to 14 pieces of manufacturing equipment.	塵埃および切屑を頻繁に除去する必要のない用途向けに設計されたサイクロン形の集塵機は，8～14個の製造装置に**使える**．
The handling system loads and unloads parts fast because it **services** both spindles at once.	このハンドリングシステムは，両方のスピンドルに同時に**使える**から，部品のロード・アンロードが早くできる．
Number of automatic machine **serviced** by one operator …….	1人のオペレータが**扱う**自動機の数は……
More than 30,000 installed and **in service**.	30,000台以上が据付けられ，目下**実用中**．
Elevators can be placed **into service**	エレベータは，乗客の動きが

during peak passenger activity and removed **from service** when desired.	ピークの間**使うように**取付け，取外したいときは撤去することもできる．
The first car／passenger ferries **came into service** in the late 1950 s.	最初の車・乗客用フェリーは，1950年代の後期に**使われるようになった**．

share: to use or possess or benefit from (a thing) jointly with others	**共用する**

Because pulsation light source **shares** the same path as the IR optics, it can't become misaligned.	脈動光源はIR光学装置と同じ通路を**共用している**から，アライメントが狂うことはない．
All batteries **used** for UPSs (uninterrupted power source system) service **share** common elements.	UPS用に**使われる**電池はすべて，共通の要素部品を**共用している**．

take: to make use of	**使う，利用する**

Taking an approximate method, ……	近似法を**使って**，……
Productivity **is** generally **taken** to mean the average output per man-hour.	生産性は，一般に，マン・アワー当りの平均生産高を意味するように**使われる**．

take advantage of: to make use of	**使う，利用する，活用する**

Cooling systems are of two types――that depending on thermo-syphonic flow, in which **advantage is taken of** the fact that hot water tends to rise and thus induces a flow in the water-circulation system, and that in which a mechanical pump positively forces water round the	冷却方式には2形式ある――熱サイフォン流れによるものは熱い水が昇ろうとすることを**利用して**，これで水循環系に流れを誘起する．また，機械的ポンプが確実に水をウォータジャケット，シリンダヘッドとラジエ

water jacket, cylinder head and radiator.

ータに強制的循環させるものもある．

Advantage has been taken of the microprocessor to integrate a number of very useful auxiliary functions. These result in the system being self-checking and also facilitate servicing.

幾多の有益な補助機能を一括できるようにマイクロプロセッサを**利用した**．これらによって，セルフ・チェックするシステムになり，また使いやすくなった．

use: to cause to act or serve for a purpose or as an instrument or as material for consumption

使う，役立てる

This type of machine **uses** single-point tools.

この種の機械は，1刃バイトを**使う**．

Cutting fluid **is used**.

切削液を**使う**．

There are many dressing methods available, but whichever **is used**, it should **employ** a rigid device.

利用可能なドレッシング方法は数多くあるが，どんな方法を**使っても**，頑丈な装置を**使う**ようにすること．

Files are tools that anyone in metal work will **use**. Often, through lack of knowledge, these tools **are misused**.

ヤスリは，誰もが金属加工に**使う**道具である．知識不足のため，この道具は**誤った使い方をされる**ことがよくある．

Martensitic stainless steels **are used** extensively by the cutlery industry.

マルテンサイト・ステンレスは，刃物工業で広く**使っている**．

Never use a wrench on moving machinery.

スパナは決して動いている機械に，**使わない**こと．

Use the correct cutting oil on the tap

ネジを切るとき，タップに適

when cutting thread.

正な切削油**を使う**こと．

On some critical components, where subsequent processing does not remove undesirable residual stresses, low-stress grinding **is** sometimes **used**.

重要な部品には，ものによりつぎの工程で望ましくない残留応力を除去しない場合，時に低応力研削法**が使われる．**

The unit **can be used** on NC or standard lathes. The tool **utilizes** a double-edged system that turns outside diameters, bores inside diameters and faces without loss of time for indexing turret position.

この装置は，NC または普通の旋盤に**使える．**工具には，タレット位置をインデックスする時間のロスがなく，外径を旋削，内径を中ぐりし，かつ面削りできる両刃方式を**使っている．**

In a similar application, a major automotive company **used** the ball burnishing process on the inside diameter of helical pinion gears and blanks. The previous method **utilized** conventional cutting and／or roll forming plus boring and honing.

似たような用途で，大手自動車会社は，ヘリカル・ピニオンギアおよびブランクの内径加工にボール・バニシング法**を使った．**それ以前の方法は，普通の切削および（あるいは）ロール成形に加えて，中ぐりおよびホーニング**を使った．**

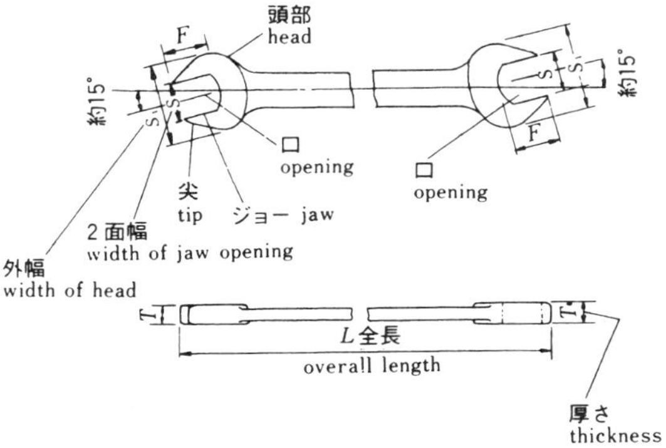

両口スパナ
(double ended spanner)

Steel shapes **are used** in bridge and heavy machinery.

型鋼は，橋および重機械に**使われる．**

This grade **is** most commonly **used** in cutting of steel.

この等級が，鋼の切削に最も普通に**使われる．**

This tool **is used** in screw machine work.

このバイトは，ネジ切り機械の作業に**使われる．**

The model test **is used** in the design of ships.

模型試験は，船の設計に**使われる．**

Bearings **used in** very low-speed applications where relative velocity between shaft and bearing is less than 10 fpm …….

軸と軸受の相対速度が10fpm以下の，ごく低速の用途に**用いられる**軸受は…….

Using a shaper **in** these situations would be economical, particularly, if the batch size is large.

これらの状況に型削り盤を**使う**ことは，とりわけもしバッチが大きいなら，経済的であろう．

Cold-drawn bar having a diameter of $1\frac{1}{4}$ inches **were used** for tests.

試験には，直径$1\frac{1}{4}$インチの冷間引抜き棒材を**使った．**

It **is** also widely **used** for manifolds and special tube fabrications, and the portable unit is beginning to **be employed** extensively in the plumbing industry.

これはまた，マニホールドおよび特殊なチューブ製作に広く**使われて**いて，このポータブル装置は配管業界で広く**使われ**始めている．

Precision, preloaded ball or roller bearings **are** widely **used** for wheel and workhead spindles, and hydrodynamic bearings **are available** on some grinders. Hydrostatic bearings **are** sometimes **used** for ways and spindles to provide increased stiffness for greater accuracy or heavier

精密級予圧玉軸受またはコロ軸受は，砥石および主軸台の軸に広く**使用されている．**また，動圧軸受も研削機によっては**利用できる．**ときには静圧軸受が，より高精度，より重荷重用に剛性を上げるために，案内面およ

duty operations. Direct drives, with the motor rotors an integrated part of the spindles, **are used** on some machines. When belt drives **are used**, the motors should be equipped with vibration-damping mountings.	び主軸に**使われる**．直結駆動装置（スピンドルと一体部品のモータ回転子付）は，機械によっては**使われている**．ベルト駆動装置を**使う**ときには，モータは振動緩衝取付け台を装備すべきである．
This kind of material can **be used** for gears to run quietly at high speed.	この種の材料は，歯車を高速で静かに運転するために**使う**ことができる．
Worm gearing **is used** for obtaining large speed reductions.	ウォーム歯車装置は，大きく減速をするのに**使われる**．
Cutting oils **used** in reaming and similar those **used** for drilling holes are ……	リーマ加工をする場合に**使われる**切削油や，穴明けするのに**使われる**類似のものは……
Chart can **be used** for determination of flow rate, minimum operating temperature, allowable distance, and smallest usable size.	チャートは流速，最小運転温度，許容距離および使用可能の最小寸法を決めるのに**使う**ことができる．
Use M2D5 rivets for thickness 0.040 in. and under.	厚さ0.040インチとそれ以下には，このM2D5リベット**を使う**．
Use both hands **to** turn the theading die.	ネジ切りダイスを回わすのに，両手を**使う**．
After wipping the table clean, **use** your hand **to** feel nicks or burrs. If you find any, **use** a honig stone **to** remove them.	テーブルをきれいに拭ってから，手を**使って**小さいノッチやバリを触感する．何か見付けたら，ホーニング砥石を**使って**それを除去する．

The operation can **be used to** enlarge an existing hole in the workpiece.

この作業は，工作物の既存の穴を拡大するために**使う**ことができる．

SOLDER ―― Generic name **used** to designate a number of fusible alloys which are conveniently **employed** for joining metals together.

はんだ――金属を互いに結合するために普通**使われ**，多くの可溶合金を表示するために**使われる**総称である．

The unit **can be used** with quick change tooling.

このユニットは，クイック・チェンジツーリングと一緒に**使える**．

To make an accurate hole it is necessary to **use** reamer with adequate support for the cutting edges, adjustable reamer may not be adequate.

正確な穴をあけるには，切刃を適切に支えてリーマを**使う**ことが必要で，調整リーマは適当でない．

Pliers should **never be used** as a substitute for a wrench, as the nut or bolt head will be permanently deformed by the serrations in the plier jaws and the wrench will no longer fit properly.

プライヤは，スパナの代わりとして，**決して使わないこと**．ナットやボルト頭がプライヤのジョーにある鋸歯形状によって永久変形し，スパナがぴったりはまらなくなるから．

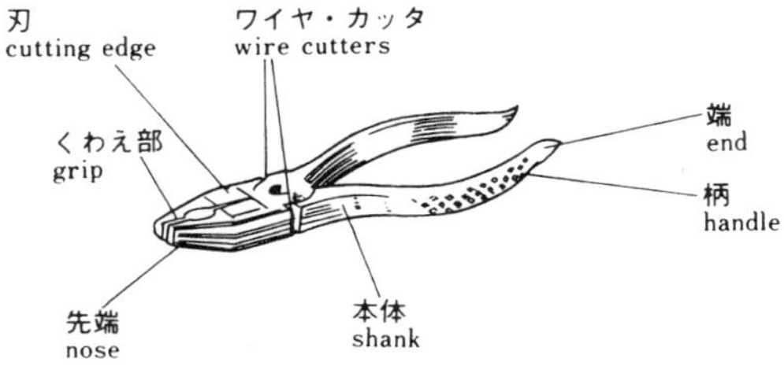

プライヤ〔ペンチ〕
(pliers with cutters, line man's pliers, cutting pliers)

Machined top surface can **be used** as work table.

切削加工した上部表面は，作業台として**使う**ことができる.

Solid lubricants **have been used** in the aircraft industry as a means of friction reduction and on forming dies, bearings, interference fit fastener systems and other mechanical systems.

固体潤滑剤は，摩擦低減の手段として航空機工業で，また成形ダイス，軸受，締まりばめ締結方式などの機械システムに**使われている.**

The most frequently **used** tool on a vertical milling machine is the end mill.

立フライス盤に最もよく**使われる**ツールは，エンドミルである.

Pivot drills are probably the most widely **used** type for producing small holes.

ピボット・ドリルは，小さい穴を作るために，たぶん最も広く**使われている**，形式である.

The most widely **used** in practice is high-speed steel.

実際に，最も多く**使われているの**は，高速度鋼である.

The dressing tools **used** should be kept sharp and properly formed.

使うドレッシング・ツールは，いつも鋭くかつ適切に成形されていること.

See figure 1-7 for required lubrication points, lubricant **to be used** and lubricating intervals.

潤滑を必要とする個所，**使う**潤滑剤および給油間隔については，図1-7を見よ.

The work is held on the machine table **using** the T slots provided.

工作物は，既設のT溝を**使って**機械テーブルに保持する.

External threading **using** dies can be carried out on turret lathes and special screw-cutting machines.

ダイスを**使って**の雄ネジ切りは，タレット旋盤および特殊ネジ切り盤ですることができる.

By **using** just one standard program a

たった1つの標準プログラム

family of parts can be machined merely by selecting various switches.

を**使って**，違うスイッチを選ぶだけで，同種の部品を切削加工できる．

……**use** under these conditions.

これらの条件で**使う**．

Coatings had no benefitical effect when **used** in a production environment.

生産環境で**使った**ときには，被覆は有益な効果がなかった．

The machine can **be used** in open workshops without the necessity of a dark room facility.

この機械は暗室設備の必要がなく，開放された職場で**使う**ことができる．

So useful and so versatile are power squaring shears that they **are used** by 42 of the 44 group in metal work industry.

この動力剪断機は，それほど有益かつ汎用性があるので，金属加工業界で44のグループのうちの42が**使っている**．

Screw threads **are used** for the purpose of fastening a screw and bolts and for the transmission of motion, e.g., a rotating screw spindle imparts a horizontal motion to a nut mounted in it.

ネジは，ネジとボルトを締める目的，ならびに運動の伝達，たとえば回転するネジ軸がそれに取付けたナットに水平運動を生じさせるために**使われる**．

…… according to the purpose for which they **are used**.

それらを**使う**目的によって，……

…… **is used** out of necessity rather than appearance.

……は，外観ではなく必要性から**使われる**．

The buttress thread **is used** in cases where large forces act in the longitudinal direction of the screw.

鋸歯ネジは，ネジの長手方向に大きな力が働く場合に**使う**．

Cylindrical roller bearings should therefore **be used** only where alignment can be

それゆえ，円筒コロ軸受は，アライメントを保つことができ

maintained.

る場合にだけ**使う**こと．

Angular-contact ball bearings **are used** for combined radial and thrust loads, and where precision shaft location is needed.

アンギュラ玉軸受は，ラジアルおよびスラストの複合荷重に対して，精密な軸の位置決めを必要とする場合に**使われる**．

Cutting fluids should **be used** when high speed cutters **are used**.

高速カッタを**使う**ときには，切削液を**使う**こと．

Use cutting oil when tapping steel, but cut thread in bronze dry.

鋼をネジ立てするときには切削油を**使う**．ただし，青銅は切削油なしでネジを切る．

Many drill heads **make use of** universal joints for multispindle drilling.

ドリル・ヘッドの多くは，多軸穴明けに自在継手を**利用している**．

In 1965 a water-cooled ruby laser developed by Raytheon Corp. **was put into use** at a Western Electric to drill diamond wire-drawing dies range from 0.001 to 0.050″.

1965年に，R社が開発した水冷ルビー・レーザは，WE社が0.001～0.050インチのダイヤモンド製線引きダイスの穴明けに**使い出した**．

They **are put to use** in turning, boring or milling nonferrous metal and plastics.

それらは，非金属およびプラスチックの旋削，中ぐり，フライス削りに**使われる**．

This system is gradually falling into **disuse** in favour of that …….

このシステムは，……のために，次第に**使われなくなりつつある**．

Laps **are in** common **use** on fine work.

ラップ盤は，精密加工に普通**使われている**．

It is important to check that the switch is in the 'off' position when the machine **is**

このスイッチは，機械を**使っていない**ときには「断」の位置

not in use.

にあることを確めることが大切である．

Center drills **are** the most rigid **in** common **use** because of their short flute lengths (usually about 4 times the diameter) and oversize shank diameters.

センタドリルは，溝の長さが短かく（普通，直径の約4倍），シャンク直径も太いので，普通**使われているもの**の中で最も剛性が高い．

Thus a computer **is of use** throughout the duration of the project.

だから，コンピュータは，このプロジェクトの期間を通して**有用である**．

utilize: to use, to find a use for　　**使う，活用する，利用する**

Utilize standard components as much as possible.

できるだけ，標準部品を**使う**．

Utilize standard preshaped workpiece, if possible.

できるなら，標準のあらかじめ成形した工作物を**利用する**．

Utilize raw material in the standard forms supplied.

支給された標準形状の原材料を**活用する**．

Utilize a table of figure.

数表を**利用する**．

Each S machine (with the exception of the 12) **utilizes** the same construction features.

S機はそれぞれ(12を除いて)同じ構造様式を**使っている**．

This crankshaft milling machine from K **utilizes** multiple tools so that cycle time for typical cranks is 90 seconds or less.

K社のクランク軸フライス盤は，代表的クランクのサイクル・タイムが90秒以下になるように，複合ツールを**使っている**．

The II-25 **utilizes** #300 spindles and the

II-25は#300スピンドルを**使**

II-30 and II-30L machines **use** #400 spindles.

い，II-30とII-30Lは#400を**使っている**.

The original concept of V-flange tooling was to **utilize** all the advantage of standardizing on a one-shank configuration …… lower inventories, fewer program variations, lower overall tooling costs because tools can **be used** on more than one machine, etc.

Vフランジ・ツーリングのもともとの考え方は，1シャンク構成の標準化で得られるすべての利点——在庫を少なく，プログラムの種類を少なく，ツールを複数機械に**使う**ことができるので全体のツーリング・コストが下がる，——など**を活用する**ことであった.

……. Both carbide and ceramic tools can **be utilized**. Servodrives **are utilized** to position the toolholders.

……. 超硬およびセラミック・ツールの両方を**使う**ことができる. ツールホルダの位置決めに，サーボ駆動装置を**使っている**.

A fluid **has to be utilized** which would be compatible with brass.

黄銅と共存性のある液を**使わねばならない**.

The bearing **utilized** by G & L has a 13″ ID, a 20″ OD, and is 1.75″ high.

G&L社が**使っている**軸受は，内径13インチ，外径20インチ，高さ1.75インチである.

Manual labor would **be employed** for associated component placement and assembly sequences. In contrast, the existing line **utilized** manpower to load, unload, and transfer workpieces.

これに伴う部品の取付けおよび組立工程には，人手が**使われる**ことになろう. これに対し，現在のラインは工作物の積卸しと搬送に人力を**使っている**.

For the remaining three 2 ¼″ holes, the manufacturer drills with the 1 ¾″ insert drill **utilized** for the other holes, then sizes them with a bigger T carbide drill. It's run

残りの3つの2 ¼インチ穴については，このメーカーは，他の穴に**使う**1 ¾インチインサート・ドリルで穴明けして，T超

at 1,000 rpm with a feed rate of 0.002″. Predrilling is necessary in this case because the machining center on which the drill **is employed** lacks the horsepower to drill a 2 ¼″ hole from the solid.

硬ドリルでより大きな寸法を出している．このドリルを**使っている**マシニングセンタは，ムクのものに2¼インチの穴をあけるには馬力が足りないので，あらかじめ穴明けしておくことが必要である．

On the larger types 150 and 500 machines separate tooling **is utilized** for the cutting of the pilot hole and collar formation.

これより大きな150および500形は，パイロット孔の切削およびカラーの形成に，別のツーリングを**使う**．

A 12″ power operated chuck **is utilized** on a L turning machine.

L旋削盤には，12インチパワーのチャックが**使われている**．

This principle **is utilized** in the cam lever or rolling lever.

この原理は，カム・レバーあるいは回転レバーに**使われる**．

If the shaft has external or internal threads, they may **be utilized** when mounting bearings.

もし軸に雄ネジまたは雌ネジがあれば，軸受を取付けるときに，それを**利用**できる．

It is necessary to **utilize** the lubricant as a means of dissipating heat.

潤滑剤は，放熱の手段として**使う**ことが必要である．

The control is easily switched from one form of positioning to the other **utilizing** the CRT and the alpha-numeric keyboard.

CRTおよび英字・数字キーボードを**使って**，制御方式をある位置決め形式から他のものに楽に切換えられる．

With the **utilization** of these natural materials, ……

これら天然の材料を**利用**して，…….

with : using as an instrument or means　　**使って，～で**

Hit it **with** hammer.

ハンマー**で**叩く．

Preheat sleeve **with** a suitable heat lamp.

適当な加熱ランプ**を使って**，スリーブを予熱する．

When the cut has started, apply cutting oil to the workpiece and die and start turning the die stock **with** both hands.

切削を始めたら，工作物とダイスに切削油を塗って，ダイス・ストックを両手**で**回わし始める．

When a lazer is used to drill a hole, it is usually used in the pulsed mode, which in effect resembles a woodpecking operation **with** a conventional drill.

レーザを穴明けに使うときは，普通パルス・モードで使われる．これは，普通のドリル**を使った**つつき作業に似ている．

Higher cutting speeds can be used **with** cutting fluids.

切削液を**使う**ことで，高速度切削ができる．

Turning the blade of a screwdriver **with** an adjustable spanner.

可調整スパナ**を使って**，ネジ回しの刃を回わす．

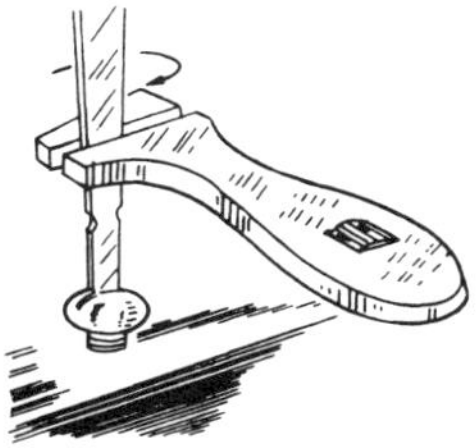

つかむ，取る，にぎる，つまみ上げる

chuck : the part of a lathe that grip the drill : the part of a drill that holds the bit	**チャック，つかむ**

Chuck the tool position with reference to the work by using a center gage. If necessary, realign the tool.

センタ・ゲージを使って，工作物に対して正しいツール姿勢に**つかむ**．必要なら，ツールのアライメントを直す．

The versatile A grips parts ranging in diameter from 1½″ to 12″ and, depending on loader configuration **chucking** parts weighing up to 250 pounds and shafts up to 500 pounds.

汎用なAは，直径で1½インチから12インチまでの範囲の部品をつかむことができ，ローダの構成にもよるが，部品は250ポンド，また軸物は500ポンドまでを**つかむ**ことができる．

Position the steady rest near the end of the shaft with the other end lightly **chucked** in a three-or four-jaw chuck.

軸の一端を三ツ爪または四ツ爪チャックで軽く**チャック**し，固定振れ止めを軸端近くに位置決めする．

Rough and finish in same **chucking**.

同じ**チャッキング**で，荒および仕上げ加工する．

〈用　語　例〉
for chuck work　チャック作業用　磁気チャック
rectangular magnetic chuck　角形　to rechuck　つかみ変える

clutch : to grasp tightly	**しっかりつかむ**

The annulus is dog-**clutched** to the driven shaft.

このアンニュラス（帯環）は，駆動軸にドッグ式**クラッチ**結合

である．

grab: to grasp suddenly, to take something greedily	**つかむ**

A robot **grabs** a part, puts it into a machining center, removes it from the machining center, and puts it into a gage.

ロボットは，部品を**つかみ**，それをマシニングセンタに置き，マシニングセンタから取出し，そしてゲージの中に置く．

That man **grabbed** me by the sleeve and pulled it with a jerk.

その男は，私の袖を**つかんで**，ぐいと引いた．

Grabbing force is adjustable to a maximum of 5,000lbs.

把握力は，最大5,000lbまで調整できる．

grasp: to seize and hold with the hand; understand	**つかむ，把握する**

Spanner.——An implement used for obtaining a firm **grasp** on a nut for the purpose of its manipulation.

スパナ——ナットを操作するとき，ナットをしっかり**つかむ**ために使う道具．

Before this is done, however the slave tool **is grasped** and pulled up to properly position the transfer shaft to remove the preload from the transfer shaft bearing.

その前に，スレーブ・ツールは，トランスファ軸がトランスファ軸軸受から予圧を取り去るのに適した位置となるまで，**つかんで**引き上げられる．

Either from your manual or from a book you will **have grasped** the rudiments of the language that you plan to use for programming.

取扱い説明書か本で，プログラミングに使おうと考えている言語の基本はすでに**理解**しているでしょう．

grip: to take a firm hold of the part of a tool or machine etc.

つかむ，にぎる，把持する

……without **regripping** the workpiece.

工作物を**つかみ直さ**ずに，……．

The 5-axis design features a dual arm which **grips** chucking parts up to 12″ in diamter and weighing up to 250lbs. each.

この5軸設計の特徴は，直径12インチ，重さ250lbまでの把持部品をそれぞれ**つかむ**2本腕である．

Different 'hand' may be inserted at the 'wrist' end of the arm, for **gripping** different objects or for scooping up powders and liquids.

違う物体を**つかん**だりあるいは粉体や液体をすくうとき，異なるハンドを，腕のリスト端に挿入できる．

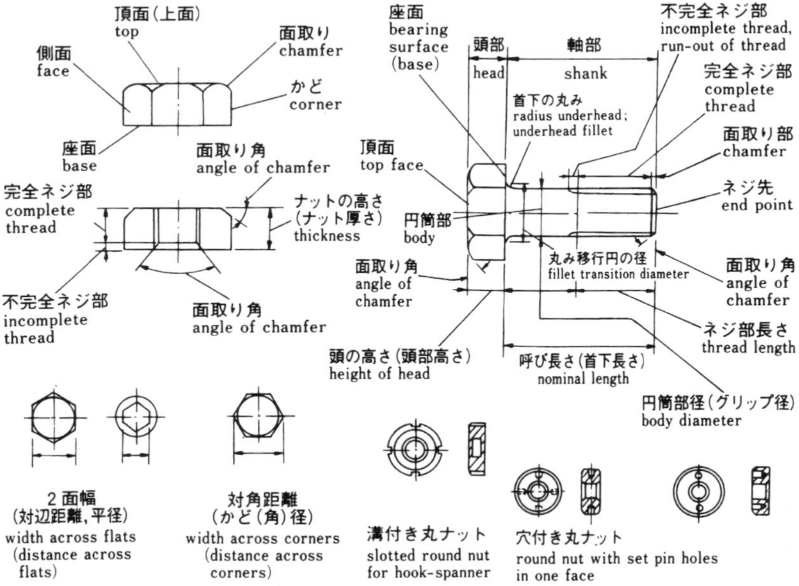

Drill jig, the M, is ideal where accurate drilling of irregularly shaped parts is required. It is fully adjustable and has the capacity to **grip** any shape securely.

穴明け治具Mは，不規則な形状部品に正確な穴明けが必要な場合にぴったりである．これは完全に調整可能で，どんな形のものも確実に**把持**できる．

A test can be made, after jacking up the front wheel of the car, by **gripping** the tyre vertically above and below the hub, and attempting to lock the wheel.

テストは，車の前輪をジャッキで上げてから，ハブの縦方向の上と下でタイヤを**つかんで**車輪をゆすって行なわれる．

Box spanner……A non-adjustable spanner which slips over a nut, **gripping** all its sides at the same time.

ボックス・スパナ――ナットの上から滑り込ませる非可調整スパナで，ナットの全側面を同時に**つかむ**．

A cylinder clamp powered by an air-hydraulic pump **grips** end of tube during bending, thus preventing inaccurate bends and saving valuable time.

空・油圧ポンプ作動のシリンダ・クランプは，曲げ作業中にチューブの端を**把持する**．これで不正確な曲りを防ぎ，貴重な時間を節約できる．

Chuck jaws **grip** on the OD of the part.

チャック・ジョーは部品の外径を**つかむ**．

They are held with standard spring-type collets that **grip on** the threads or shanks.

それらは，ネジまたはシャンクを**つかむ**標準のバネ形コレットで，保持する．

The workpiece **is gripped** at one end by a chuck mounted on the end of the main spindle of the machine and is supported at the other end by a center in the tailstock.

部品の一方の端は機械の主軸端部に取付けたチャックで**つかみ**，他端は心押し台のセンタで支持する．

When removing a difficult split pin it should **be gripped with** the cutting edge of

割ピンの取外しが困難なときには，プライヤの刃で**つかむ**こ

the pliers.

と．

The diaphragm chucks **grip** precision parts with controlled force.

ダイヤフラム・チャックは，力を加減して精密部品を**把持する**．

The tool is rotated by hand, the workpiece **being gripped** in a vise. Alternatively, the workpiece may **be gripped** in the chuck of a lathe and rotated, while the tap or die is guided by the tailstock.

切削工具は手で回わし，工作物をバイスで**つかむ**．その代わりに，タップまたはダイスを心押し台で案内して，工作物を旋盤のチャックで**つか**んで，回わしてもよい．

Nominal grinding capacity is 76mm diameter and 114mm long, the maximum diameter which may be ground is dependent on the external dimensions of the component which can **be** effectively **gripped** in the workholding fixture within a 254mm diameter.

呼称研削能力は直径76mm，長さ114mmで，研削できる最大径は，工作物保持取付け具が実際に**つかむ**ことのできる部品の外部寸法（直径254mm以内）による．

A-system collets provide more **gripping** power than any other tapered collet on the market. Plus, each collet can handle an amazing range of tool sizes.

A方式のコレットは，市場にある他のテーパ・コレットよりも大きい**把持**力がある．加えて，コレットはいずれも驚くほど広汎な寸法のツールを扱うことができる．

The jaws **have** either ID or OD **gripping ability** depending on the finger design.

チャック爪は，そのフィンガの設計によって，内径，外径のいずれかを**把持することができる**．

After rough and finish turning, the chuck jaws retract, exposing the **gripping** area for finish machining. The part is complete

荒および仕上げ旋削すると，チャックの爪は引込んで，仕上げ切削できるように，**把持して**

in a single operation with greater productivity and profit.

いた部分を露出する．部品は，たった1回の作業で完成し，生産性および利益も大きい．

The **gripper** has two heads, for simultaneous loading and unloading operations. Each head has two independently programmable, V-type jaws.

グリッパには，ローディングおよびアンローディング作業が同時にできるように，2つのヘッドがある．各ヘッドには，別々のプログラミング可能な2つのV形ジョーが付いている．

Shafts can be loaded by using the two arms in combination. The universal **gripper** jaws are adjustable, and will accept parts 1.5″ in diameter.

軸は，2つのアームを一緒に使うことによって，ローディングすることができる．万能**グリッパ**・ジョーは調整式で，直径1.5インチの部品を受け入れる．

The H **grips** the paper between a hard rubber pinch wheel and a metal surface covered with aluminum oxide grit. The accuracy of this alignment allows the plotter to operate with both ends of the paper free.

Hは，紙を硬製硬質ゴムのピンチ車と酸化アルミニウム粒子で覆われた金属表面の間に**把持する**．このアライメントは正確なので，紙の両端を抑えずに，プロッタを操作することができる．

The two fabric discs **are gripped** under spring pressure between the pressure plate and the flywheel.

2つの織物製円板は，圧板とフライホイールの間のバネ圧で**把持されている**．

It is true that plastic cannot **grip** a coating as tight as metal can.

プラスチックは，金属のようにしっかりと被覆物を**把持**できないことは事実である．

The wheels 'bounce' to a certain extent and therefore **do not obtain** quite so good **a grip on** the road surface.

車輪はある程度バウンドするため，路面をそれほど良くは**つかめない**．

The **friction grip** to these parts in their seating is ample to prevent rotation.

座の部分におけるこれらの部品に対する**摩擦把持**で，まったく回転しないようになる．

Diamond knurling is used to improve the appearance of the part and to provide a good **gripping** surface for levers and tool handles.

綾目ローレット切りは，部品の見てくれを良くするため，ならびにレバーおよびツール・ハンドル用に良い**把持**表面を供するために使われる．

〈用　語　例〉

friction grip bolt　摩擦接合用ボルト
gripper　グリッパ，つかみ（やっとこ）
grip tong (upsetter)　グリップ端
handle grip　握り

hook：a bent or curved piece of metal etc. for catching hold or for hanging thing on
：to grasp or catch with a hook, to fasten with a hook or hooks

（鈎で）ひっかけてつかむ

Removal of the restrictor pin is effected by **hooking** a length of wire through the hole in the end.

制止ピンは，適当な長さのワイヤを，ピン端の穴に通して**ひっかけ**て取外す．

Hook a 50-gram weight **to** the loop at each end.

各端のループに，50グラムの重錘を**ひっかける**．

For testing an engine, one end of the chain is **hooked** on to a convenient point of the engine, and the other portion laid on all the plug terminals except the cylinder that is desired to test.

エンジンの試験には，チェーンの一端をエンジンの適当なところに**ひっかけ**，その他の部分を試験しようとするシリンダ以外のプラグ端子全部に乗せる．

When no pointer is fitted it is often possible to judge the tension of each adjusting nut by engaging one end of a

指針がついていないときには，スパナの一端をナットに掛け，図307のように他端にバネ秤り

spanner with the nut and **hooking** the spring balance to the other end, as shown in Fig. 307.

をひっかけることによって各調整ナットの張力を判定できる場合が多い．

pick up: to lift or take up　　**つまみ上げる，取上げる**

As one gripper **picks up** an unground part, the other gripper deposits a finish ground part on the conveyor. The carriage then moves the grippers to the grinding station.

一方のグリッパが研削していない部品を**取上げる**と，他のグリッパが仕上げ研削した部品をコンベヤに置く．つぎに，キャリッジが，グリッパを研削ステーションまで動かす．

An attendant puts an unworked part in a nest. The robot **picks up** the part and automatically…….

機械についている人が未加工品を部品巣箱に置く．ロボットはその部品を**つまみ上げ**て，自動的に…….

Do not **pick up** a cathode-ray tube by the neck, or let it fall even a distance of 1 or 2 inches.

ブラウン管は首のところを**つまみ上げ**たり，たとえ１～２インチの高さでも落とすようなことがあってはならない．

The elevating mechanism centers the blank, positioning it to **be picked-up** by the V-type jaws of the overhead gripper.

この昇降機構は素材の中心を出し，オーバーヘッド・グリッパのＶ形ジョーで**つまみ上げられる**ように位置決めする．

In automatic mode of operation, an operator fills a magazine chute with blanked parts, where individual parts are continuously shuttled from the bottom of the stack to a nested **pick up** position.

自動方式の作業では，作業者がマガジン・シュートに打抜き部品を充塡する．ここで部品は個別に，架の底から巣箱になっている**ピックアップ**位置に，連続的に繰出される．

You have misused the “file-handling”

「ファイル取扱い」編の使い

part and so not **picked up** any data.

方を誤っていたら，データはまったく**取出せない**.

snatch : to seize quickly of eagerly — **ぐっとつかむ**

Do not **snatch** the gear lever into position, but do so with a slow, firm, and progressive movement.

所定位置にギアレバーを**ぐいとつかんで**入れずに，ゆっくりしっかり漸次動かすようにすること.

take : to lay hold of, to capture — **取る**

Take a small quantity in a test tube.

試験管に，少量**取る**.

When this is the case care should be taken to remove all external locking parts. The selector locking springs and balls may have to **be taken** out, for instance, after unscrewing the plugs, before the lid is lifted.

こういうときには，外側の固定部品を全部取外すように注意すること．ときには，ふたを持上げる前に，プラグをねじ戻してから，セレクタ固定バネと球を**取**出さなければならないこともある.

A robot picks up a part, moves it, put it in a fixture, **takes** it out of the fixture, and moves it again.

ロボットは部品を取上げ，移動し，そして取付け具に置き，取付け具から**取り**，ふたたび動かす.

trap : to catch or hold in a trap — **つかまえる，捕捉する**

The cup shaped slinger above the bearing **traps** oil as it comes to rest.

軸受が停止すると，軸受上部のカップ状油切りが油を**捕捉する**.

The cage would normally be the first component to fail, and a ball or roller may **be trapped**, resulting in complete bearing seizure.

通常，最初にだめになる構成部品は保持器で，玉やコロがこれに**つかまり**，完全に軸受が焼付くようになる.

突 切 る

cutting off：cut the right angle with tool　**突切る**

A variety of **cut off** tool holders are used to **cut off** or make groove in workpieces.

工作物を**突切る**，あるいは溝を作るのに，いろいろな**突切り**バイトホルダが使われる．

Turning between centers has some disadvantages. A workpiece cannot be **cut off** with a parting tool while being supported between centers, as this will bend and break the parting tool and ruin the workpiece.

センタ支持旋削にも都合の悪いことがある．センタ間に支持したまま，突切り工具で工作物を**突切れ**ない．これでは，突切り工具が曲って破損し，工作物も駄目になるからである．

Parting or **cut off** tools are often used for necking or undercutting, but their main function is cutting off material to a correct length.

突切りや**切断**用工具は，逃げ溝切りやアンダーカットによく使われるが，その主たる機能は材料を正しい長さに切断するこ

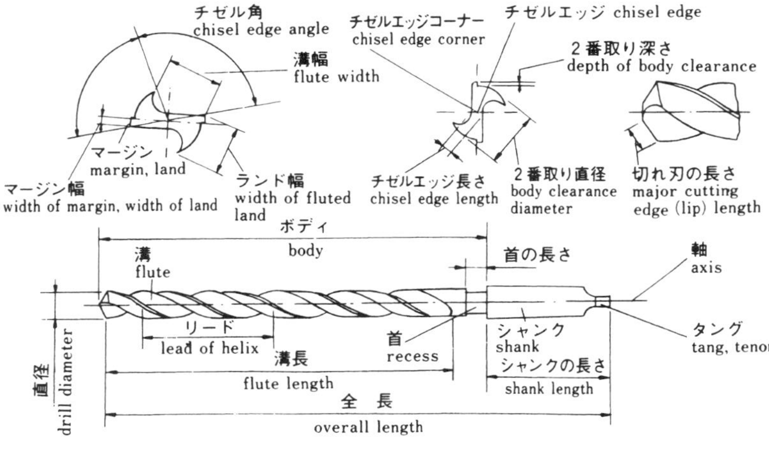

とである．

〈用　語　例〉

bar cut-off machine　棒鋼切断機　travelling cut-off saw　移動丸鋸
longitudinal feed cut-off tool　縦式横切り盤
突切り工具

part: to separate or divide, to cause to do this　**突切る**

When deep **parting** difficult material, extend the cutting tool from the holder a short distance and **part** to that distance. Then back off the cross slide and extend a bit further; **part** to that depth. Repeat the process until the center is reached.

深い**突切り**が困難な材料のときは，バイトをホルダから短かい距離出して，その距離まで**突切る**．それから，横送り台を戻して，さらにちょっと出す；その深さまで**突切る**．このやり方を，中心に達するまで繰返す．

Parting alloy steels and other metals is sometimes difficult, and step **parting** may help in these cases.

合金鋼その他の金属は，ときに**突切ること**が困難であるが，そういう場合には段階的**突切り**が役に立つ．

Work should not extend very far from the chuck when **parting** or grooving, and **no parting** should **be done** in the middle of the workpiece or at the end near the dead center.

突切りや溝切りするときには，工作物をチャックから遠く出さない．そして工作物の中央あるいは止まりセンタに近い端で**突切りしないこと**．

All **parting** and grooving tools have tendency to chatter, therefore any setup must be rigid as possible.

突切り工具および溝切り工具は，すべてガタつく傾向がある．したがって，取付けはできるだけ強固でなければならない．

Lathe tools are ofen specially ground as **parting** tools for small or delicate **parting** jobs.

旋盤バイトは，小物や繊弱なものの**突切り**作業用の**突切り**バイトとして，特別に研削することがよくある．

〈用　語　例〉

parting tool　　突切りバイト

sever: to cut or break off from a whole, to separate	**突切る**

The rotary slotting attachment is used to slot or mill the out end of a workpiece, before it **is severed** from the bar.

この回転立削り装置は，工作物を棒材から**突切る**前に，その外側の端を立削りあるいはフライス削りするのに使われる．

つく（創）る（創造，考案）

build: to construct by putting parts or material together	**つくる，組立てる**

In 1779 James Watt **built** the first successful steam engine.

1779年，ジェームス・ワットは，最初の蒸気機関を首尾よく**つくった**．

Manufactured by the George S. Lincoln Company in Hartford, it became famous under the name of the Lincoln Miller. Over 150,000 **were built**, and some worked continuously for more than 70 years.

Hartford にあるＬ社が**製作した**これは，Ｌフライス盤の名で有名になった．15万台以上が**つくられ**，70年以上連続して働いたものもあった．

The first completely automatic turret lathe for turning metal was designed and **built** by Christopher Miner Spencer.

金属旋削用の最初の完全自動タレット旋盤は，C.M.スペンサーが設計，**製作した**．

The unit **is built** from modular units to meet machining and materials handling requirements.

このユニットは，切削およびマテリアル・ハンドリングの各条件に合うように，モジュラー・ユニットで**できている**．

It **is built** to a tolerance of ±0.001 inch and, in certain areas, has to be square.

これは，±0.001インチの公差に**つくり**，ある部分は直角でなければならない．

Some lathe **are built** with a leadscrew and a feed rod.

親ねじと送り軸付旋盤も**つくられている**．

Built into one of these struts is the drive shaft of the gear system.

歯車系の駆動軸は，支柱の1つに**組み込まれている**．

Built-in chip conveyor.

造り付けの切屑コンベヤ．

The LS-1740 **is built** for high metal-removal rates in turning applications.

LS-1740は，旋削作業で重切削ができるように**つくってある**．

We **build** special, high-production, single-purpose drilling machines——all custom **built** strictly suited to the application.

特殊な高生産性，専用ボール盤を**製作**——用途にぴったり適合，すべて受註**製作**．

Every **rebuilt** machine is thoroughly inspected and tested in our plant prior to shipment.

つくり直した機械はどれも，出荷に先立って，工場でくまなく検査・試験する．

construct : to make by placing parts together	（組合せて）つくる

A spring-loaded 4-ball wear tester **was constructed**.

バネ負荷4球摩耗試験機を**組立てた**．

The landing gear is of the skid type, **constructed** of tublar aluminum alloy.

着陸装置は，管状のアルミ合金で**つくった**スキッド形である．

A calibration curve **is constructed** from Table 1.

表1から，補正曲線を**つくる**．

Trial construction of machine is done actively.

機械の**試作**を精力的に行なった．

Nuclear power plants **under construction**…….

建設中の原子力発電所は……，

Coaxial cable is used for a lot of data-handling jobs. It's available in many different **constructions** for different applications and environments.

同軸ケーブルは，多くのデータの処理作業に使われる．これには，いろいろな用途および環境に合うよう，いろいろな**構造**のものがある．

The housing and bearing retaining washer of the bearing tester **are constructed** so that they act much like the shields of the shield bearings.

軸受試験機のハウジングと軸受止め座金(リテーナ・ワッシャ)は，それらがシールド軸受のシールドによく似た働きをするように，**つくられている．**

The welded-steel **construction** has six pieces of plates with the side members made in shape of an "E".

この溶接鋼**構造物**は，E形に作られた側部材と6枚の板でできている．

create : to bring into existence, to originate, to give rise to

つくる（創る）

To <u>**make**</u> hardened cutters for his shaping machine, Fellows **created** another machine.

フェローは，自分の歯車形削盤用の焼入れカッタを<u>**つくる**</u>ために，もう1つ機械を**創った．**

The first tantalum carbide-tungsten carbide cutting tool material **was created** <u>by</u> FANSTEEL in 1931.

最初のTa・WC超硬切削工具は，1931年にファンスチール<u>が</u>**創った．**

The processors **are created** <u>by</u> the operator for most NC lathes machining

このプロセッサは，ほとんどのNC旋盤，マシニング・セン

centers and punch presses.

タ，パンチプレス用に，オペレータが**作る．**

erect : to set up, to build　　**つくる（建造する）**

Frame, ——A term which in motor engineering usually refers to the light frame of steel upon which the car **is erected**.

フレーム——自動車技術では，通常，車をその上に**組立てる**鋼製の軽い枠を指す用語．

At the time of **erection**, the teeth of the bull gears for azimuth and elevation positioning were rough——about 300rms.

建設の時点で，方位および仰角位置決め用のブル・ギアの歯は粗く——約300rms——であった．

generate : to bring into existence, produce　　**つくり出す**

……**generate** hydrogen from zinc and sulfuric acid.

亜鉛と硫酸で，水素を**つくる．**

Figure 1．3(b) shows a cylindrical surface **being generated** on a work-piece by the rotation of the workpiece and the movement of the carriage along the lathe bed ; the operation is known as cylindrical turning.

図1．3(b)は，工作物の回転と旋盤のベッドに沿った往復台の運動とによって，工作物に**つくり出される**円筒面である：この作業がいわゆる円筒旋削である．

The tooth profiles of bevel gears **are generated** by tools carried on a rotating cradle and representing teeth of a crown gear or another bevel gear.

傘歯車の歯の形は，回転クレードルに取付けられ，冠傘歯車などもう一方の傘歯車の歯に代わるツールによって**創成される．**

After a hole **is generated** by the drill bit, formation pins on the D tool pull up a T-joint collar, all in a single cycle.

ドリル・ビットで穴を**つくった**あと，Dツールのフォーメーションピンが，T-ジョイントカラーを引上げるというのが1

サイクルの動作である．

A simple abrasion tester has been developed that **generates** data correlating closely with results in the field.

現場の結果とぴったり相関するデータを**つくり出す**簡単な摩耗試験機を開発した．

invent : to create by thought ; to make or origin (something that did not exist before)	**考え出す，発明，案出**

The trasister **was invented** in 1947 at Bell Telephone Laboratories to replace the bulky glass tubes that controlled and amplified electric currents in early TVs and computers such as ENIAC.

トランジスタは，1947年ベル電話研究所で，初期のTVおよびENIACのような，コンピュータの電流を制御および増幅するための嵩ばったガラス真空管に置換えるために，**発明された**．

In the cardan gears (**invented** by Cardano) shown in Fig. 6 , the inner wheel has half the radius of the outer.

図6のカルダン歯車装置（カルダノの**発明**）では，内輪は外輪の半径の半分である．

make : to cause to exist, to produce, to create	**つくる**

A shock wave which a jet plane flying at supersonic speed **makes** in the air……

超音速で飛行中のジェット機が空気中に**つくる**衝撃波は……

produce : to come forth, to originate	**つくる**

A smooth surface can **be produced**.

滑らかな表面を**つくり出す**ことができる．

The axis of the cutter is usually parallel to the axis of the workpiece, but in some cases tilting may be necessary to **produce** the helix angle of the thread.

カッタ軸は，通常工作物の軸と平行であるが，場合によっては，ネジのねじれ角を**つくる**ために，傾けることが必要になる．

つくる（構成）

formulate : made up of　　つくる

Sunicool EP, a fully synthetic coolant, **is** specifically **formulated** for high-speed machining operations requiring high cooling and lubricity properties.

純合成冷却液 Sunicool EP は，優れた冷却および潤滑性を必要とする高速切削作業用に，特別に**つくった**ものである．

This new coolant / lubricant for stamping and cutting tools **is formulated** to transfer friction heat from the tool to the metal chip.

型打ちおよび切削工具用のこの新しい冷却・潤滑剤は，摩擦熱を工具から金属切屑に伝導するように**つくられている**．

make up : to form or constitute, to put together, to prepare　　つくる，組立てる

Make up any desired dimensions by means of a thickness gauge.

厚さゲージによって，所望のどのような寸法のものでも**つくる**．

This unit **is made up** of two parts : the mechanical head which holds the workpiece and the electronic unit which controls the rotation of the spindle.

このユニットは，２つの部分で**できている**：工作物を保持する機械的ヘッドと，主軸の回転を制御する電子ユニット．

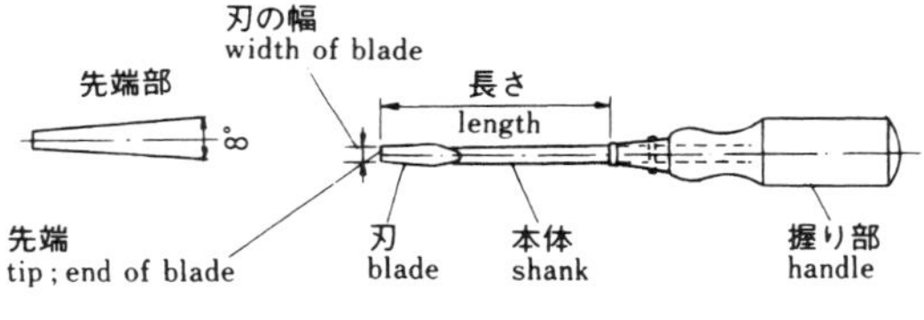

ネジ回し（screw driver）

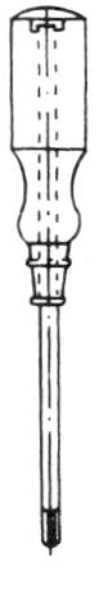

十字ネジ回し
(screw driver for cross recessed head screw, Phillips type)

つくる（作成，製作，製造，生産）

fabricate : to construct, to manufacture　　**つくる**

Fluorosilicone elastomers are easy to **fabricate**.

フロロシリコン・エラストマは，楽に**つくれる**．

The 202 and 205 prefixed parts may **be fabricated** by the operator in accordance with the information in the Service Letter.

202および205を前に付けた部品は，作業者がこのサービス・レターの資料に従って**つくって**もよい．

If shims are required as a result of this inspection, they should **be fabricated** locally from aluminum sheet material of the desired thickness.

この検査結果，もしシムが必要であれば，所望の厚さのアルミ薄板材から，現地で**つくる**こと．

The large majority of H concrete anchors **are fabricated** from 12L14 steel, though specials are sometimes **produced** from other materials such as stainless steel where extreme corrosion resistance is necessary.

H.コンクリート・アンカーの大部分は，12L14鋼で**つくられている**．きわめて耐蝕性が必要な場合にステンレス鋼が使われるように，ときには特別のものを，他の材料で**つくる**こともある．

The test shafts **are fabricated** of 17-4 P_H steel and are heat treated to a hardness of RC 45～47.

試験軸は，17-4 P_H 鋼のものを**つくり**，硬さ RC 45～47に熱処理する．

The **fabricating** technique for the filament-wound composite bushings is as follows:

フィラメントを巻いた複合ブッシュの**製作**技法はつぎの通り：

The standard Fixed Height Table is a

標準の固定高さのテーブルは

casting, while the special requirements **are fabricated** to customer specifications.	鋳物であるが，特別要求のものは顧客の仕様通りに**つくる．**
TN allows direct **fabrication** of optical components from suitable materials, using a single-crystal diamond tool and a specially **built** machine tool.	TNによって，単結晶ダイヤモンド工具と特別**製の**工作機械を使って，適切な材料から光学部品を直接**つくる**ことができる．
make: to construct or creat or prepare or form parts or form other substance	**つくる**
……**make** a 3/8-4 Acme internal thread in a cut.	1回の切削で，3/8-4アクメ雌ネジを**つくる．**
……**make** a square hole in a large solid die block.	大きなムク型材に，四角の穴を**あける．**
Slots **were** previously **made** in the tubes by milling with a 0.002″ wide cutter. This operation produced sharp burrs that had to be removed in a subsequent tumbling operation.	従来,幅0.002インチのカッタを使ったフライス加工で，チューブに溝を**つくった．** この作業では，鋭いバリができるため，つぎにタンブリング作業でそれをとり除かなければならなかった．
This division **makes** custom-made assembly machines for the automobile industry.	この部門は，自動車工業向けの受注製作組立機械を**つくっている．**
M **makes** four switch types in its G Series.	Mは，G系列4タイプのスイッチを**つくっている．**
……**is** custom **made**.	……は受注**生産である．**
Cubic-structured boron nitride (CBN), a **man-made** abrasive crystal approaching	立方晶形窒化硼素(CBN)(ダイヤモンドの硬さに近付いてい

the hardness of diamond, is being used for tool grinding.	る**人造**研摩剤結晶）が，ツールの研削に使われている．
These aluminum complex greases **are made** by our company.	このアルミニウム複合グリースは，当社が**製造した**ものである．
The tools used **are made** of special steel (tool steel), hard metals (cemented carbide alloys), oxide ceramic materials, and diamonds.	使用工具は，特殊鋼（工具鋼），超硬合金（焼結炭化物合金），酸化物セラミック材，およびダイヤモンド**製**である．
Steel propellers **are made** <u>of</u> hollow construction.	鋼のプロペラは，中空構造<u>で</u>**つくられている**．
These tools **are made** from high carbon or high speed steel.	これらの工具は，高炭素鋼またはハイス**製**である．
Massive bases **made** <u>from</u> heavy castings with thick walls and cross ribbing, or rigid weldments with sturdy reinforcements are necessary to withstand high grinding forces and maintain accuracy.	厚い壁とはすかいリブ付きの重量のある鋳物，あるいは頑丈な補強材のついた高剛性溶接構造<u>で</u>**つくった**どっしりしたベースが高研削力に耐え，かつ精度を維持するために必要である．
Make screw <u>from</u> coiled wire.	ネジをコイル状ワイヤ<u>から</u>**つくる**．
Grooves in bores **are** usually **made** <u>by</u> feeding a formed tool straight into the work.	穴の中の溝は普通，総形工具を工作物にまっすぐ送り込ん<u>で</u>**つくる**．
To **make** a part <u>by way of</u> the numerically controlled process, a drawing with accurate engineering dimensions **is made**.	数値制御方式という<u>方法で</u>部品を**つくる**ため，正確な製作寸法を記入した図面を**つくる**．

A square shoulder **is made** with a counterboring tool.

直角ショルダーは，座ぐりカッタを使って**つくる．**

Heavy duty bearings **are made** with relatively heavy outer rings, **made** from hardened and precision-ground materials.

重荷重用軸受は，焼入れ，精密研削した材料で**つくった**，比較的どっしりした外輪付きで，**つくられている．**

The cutters **are made** with various CPM grades.

カッタは，いろいろなCPM等級で**つくられている．**

The first specimen blanks for this machine **have been made** in both cutting and forming sizes from normalized SAE 1117 steel and SAE 1018 steel.

この機械用の最初の試料であるブランクは，切削および成形両方の寸法に焼なましした SAE 1117鋼および SAE1018鋼で，**つくった．**

A dent-removing tool consists of an adjustable frame and two rollers. The rollers **are made** in various contours and are interchangeable.

へこみ除去工具は，調整式の枠と２つのローラでできている．ローラは，いろいろな輪郭に**つくられていて**，互換性がある．

Because they are used for high speed, cylindrical roller bearings **are** commonly **made** in precision grade such as ABEC-5, as well as the lower grades.

高速で使われるので，円筒コロ軸受は，低等級のものだけでなく，普通 ABEC-５のような精密級で**つくられている．**

Some of these oils **are made** to be applied as an emulsion with water.

これらの油のあるものは，水とのエマルジョンとして使うように**つくられている．**

J index tables **are made** from heavy box section castings to withstand clamping pressures without distortion.

割出しテーブルは，クランプ圧に変形なく耐えられるよう，重い箱形断面の鋳物で，**つくられている．**

manufacture: to make or produce (goods) on a large scale by machinery	つくる（製造，生産）

We **manufacture** and market a complete line of lubricants.	潤滑剤の全品種を**製造**，販売している．
The saw blades **are manufactured** from the highest grade of M2 super high speed steel (DM05), coated with a special treatment for better blade life and lubricating qualities and hardened to an R_c of 63～65.	この鋸刃は，最高級のＭ２超高速度鋼で**製造されており**，刃の寿命および潤滑性を良くするために特殊な処理でコーティングし，かつ Rc63～65に焼入れしている．
Screw **is manufactured** on this type of lathe.	ネジは，この形式の旋盤で**製作する**．
Ceramic tool material must **be manufactured** in insert form.	セラミック工具材は，インサート形状で**製作**しなければならない．
The W vertical boring and turning mills **are manufactured** in both standard and CNC models.	竪形中ぐり・旋削盤は，標準形とCNC形の両方で，**製造されている**．

produce: to originate, to manufacture	つくる（製作，生産）

Produce the desired surface shape.	望み通りの表面形状を**つくる**．
Full profile inserts are available in any of the common thread forms. They **produce** a technically perfect profile (root and crest) for that particular pitch thread.	総形インサートには，普通のネジ形状ならどんなものもある．これで，特定のピッチのネジについて，技術的に完璧な輪郭形状（谷および山の頂）が**できる**．

The Eldrado Mega75 gundrilling system is capable of **producing** deep or shallow holes 0.0781″ to 1.000″ diameter and desired bottom form in one pass.

E.M75穴明けシステムは，0.0781～1.000インチ径の深穴または浅穴および所望の底形状を，1回のパスで**つくる**ことができる．

When used to machine cast iron, the taps generally **produce** at least 10,000 holes (in some applications it's a much higher number) before resharpening is necessary.

鋳鉄を切削するのに使ったとき，このタップは，一般に研ぎ直しを必要とするまでに，少なくとも10,000個の穴が**つくれる**（用途によっては，さらに多数）．

……**produce** a rectangular slot in the workpiece.

工作物に，矩形の溝を**つくる**．

A 20-TON, 65″ stroke, fully automated hydraulic broaching machine is **producing** a nonuniform external splined section on an automotive automatic transmission final drive internal gear at the rate of 160 pieces per hour.

20トン，65インチストローク全自動油圧ブローチ盤は，毎時160個の速さで，自動車の自動ミッション**最終駆動内歯車**に非対称の外スプライン部分を**つくっている**．

The milling machine is capable of **producing** a variety of large-and medium-sized parts.

このフライス盤は，各種の大および中サイズの部品を**つくる**ことができる．

Along with the millions of drill bushings **produced** annually, Welch also **makes** hundreds of thousands of different tooling components.

年間**製造する**数百万個のドリル・ブッシュとともに，Wはまた数10万のいろいろなツーリング部品を**つくっている**．

……enable the automobile and airplane to **be built** efficiently and mass **produced** economically.

……で，自動車および航空機を効率良く**つくり**，かつ経済的に大量**生産する**ことができる．

Most spur gears **are** mass **produced** on production gear making machines. Occasionally a spur gear needs to **be made** on a milling machine.

ほとんどの平歯車は，生産形歯車加工機で大量**生産される．**場合によっては，平歯車は，フライス盤で**つくる**必要もある．

Screw threads can **be produced** in various ways : by hand with the aid of such devices as a screw tap, a die plate or a screw die, or by mechanical methods which comprise turning (on a lathe), milling, rolling, pressing or casting.

ネジは，いろいろなやり方で**つくる**ことができる：ネジタップ，ダイプレートやネジダイスのような道具を用いて手で，あるいは旋削(旋盤で)，フライス削り，転造，プレス加工あるいは鋳造などの機械的方法で．

Tantalum carbide **is produced** by combining tantalum with carbon and heating to about 3,000°F.

Ta カーバイトは，Ta と C を結合して約3,000°F に加熱して，**つくる．**

The first improved tool steel **was produced** in 1868 in England by Robert Mushet.

最初の改良工具鋼は，1868年に英国で Robert Mushet が**つくった．**

With his new machine, Heald **produced** finished cylinder perfectly straight and parallel to within 0.00025 inch.

彼の新しい機械で，0.00025インチ以内に，完全に真直で平行に仕上げられた円筒を**つくった．**

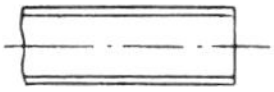
あら先
plain-sheared end, unpointed end

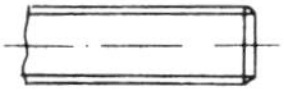
平先
flat point

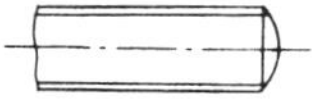
丸先
round end, round point, oval point

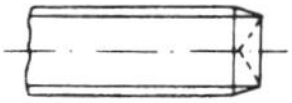
窪み先(輪先)
cop point, half point

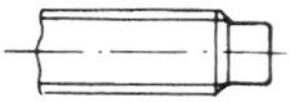
棒先
full dog point

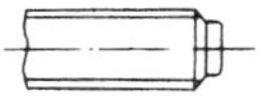
半棒先
half dog point

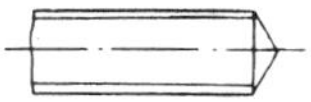
とがり先(剣先)
cone point

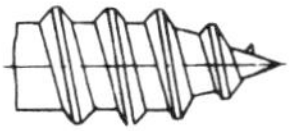
巻き先
gimlet point

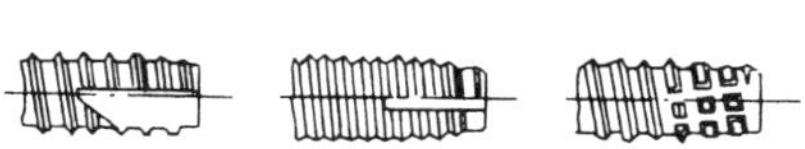
溝先(スリット) fluted point

ネジ screw

Greases **are produced** with Al, Ca bases.

Aℓ, Ca 石鹸基の，グリースを**つくる．**

Small (7mm diameter) disks of filter paper **were produced** using a paper punch.

紙打抜き工具を使って，沪過紙の小さい（直径 7 mm）円板を**つくった．**

Brackets, flanges, rails and plates of all types are just a few of the innumerable kinds of parts that can **be produced** on this remarkable new precision machining center.

あらゆる形のブラケット，フランジ，板物は，このすばらしい新形精密マシニングセンタで**製造**できる無数の種類の部品のうちのほんの 2， 3 である．

Carborundum **is produced** in the electric furnace.

カーボランダムは，電気炉で**つくる．**

Produce a certain article in gross lots, keeping each gross separate.

特定の品物はグロス・ロットで**つくり**，グロス別にする．

Produce necessities at minimum costs.

最小コストで，必要なものを**つくる．**

This final assembly-test transfer system **produces** front wheel drive automotive axle assemblies at a rate of one every three seconds.

この最終の組立・試験トランスファシステムは， 3 秒に 1 個の速さで前輪駆動の自動車アクスル組立品を**生産する．**

Cylindrical Grinder **produces** parts to tolerances in millionths.

円筒研削盤は，部品を許容差 $1/10^6$ に**つくる．**

Easy operation of the machine enables relatively unskilled operators to **produce** parts to close limits without expensive toolings.

機械の操作がやさしいから，比較的未熟な作業者でも，高価なツーリングを使わずに狭い公差に部品を**つくる**ことができる．

Our Burkland-tooled Schmid fine-blan-

精密打抜きプレスは，800トン

king presses have capacities up to 800 metric tons and can **produce** parts up to 20-in. dia. and 5／8-in. thick from coil or flat stock.

までの能力があって，直径20インチで厚さ5/8インチまでの部品を，コイルまたは平材から**つくる**ことができる．

Production of screw threads can **be accomplished** by use of taps.

タップを使うことによって，ネジを**製作する**ことができる．

Durakut cutting tools have been **in production** for over 70 years. The KSD tools **are manufactured** by Kobe Steel Ltd. of Japan and are available through Durakut International.

D切削工具は，これまで70年以上**生産されている**．KSDツールは日本のK社**製**で，D社を通じて手に入る．

turn out: to produce by work　　**（加工して）つくる**

The **production**-boasting Excell-O-Model 1UC turn／bore／contour slant bed machine **turn out** high-volume precision work.

生産自慢のE旋削・中ぐり・輪郭削りの傾斜ベッド機Excell-O-Model 1UCは，大量の精密加工品を**つくる**．

Because machines of this type generally **turn out** two parts per cycle, the production rate is about 120 parts per hour at 100 % efficiency.

この形の機械は一般に1サイクル当り2個の部品を**つくる**から，生産速度は効率100％で毎時約120個である．

Batch **manufacturing** plays an extremely important role in the nation's rate of productivity. Indeed, some 75% of the parts **manufactured** in the United States **are turned out** in lots of 50 or less.

バッチ**生産**は，国の生産性にきわめて大きな役割りを演じている．事実，米国で**製造される**部品のほぼ75%が，50あるいはそれ以下のロットで**つくられている**．

吊す，懸ける，掛ける

hang: to support or be supported from above so that lower end is free.	吊す

The instruments **are hung up** at the windows to dry.

道具は，窓に**吊して**乾かす．

In no circumstances should the gearbox be allowed to **hang** in the clutch assembly during removal or refitting, owing to the risk of distorting the driven plate, or bending the splined shaft.

どんな場合でも，歯車箱を取外しまたはふたたび取付けるときに，クラッチ機構の中に**吊下げたままにして**はならない；駆動板を歪ませたりスプライン軸を曲げる危険があるから．

suspend: to hang up, to keep from falling or sinking in air or liquid etc.	吊す

Suspend by iron bar at each of the four corners.

四隅それぞれで，鉄の棒により**吊す**．

Suspend a thin and light mirror by a fine quartz thread of about 1μm diameter.

薄くて軽い鏡を，直径約1μmの細い石英の糸で**吊す**．

Suspend the substance to be measured **with** a fine thread.

測定する物体を，細い糸で**吊す**．

A weight **is suspended** from the end of the spindle.

錘りを，主軸の端から**吊す**．

The slurry pump is equipped with an agitator to keep solids **in suspension** prior to pumping.

スラリーポンプは，ポンピングに先立って，固体が**懸架している**ように，攪拌機を備えている．

〈用　語　例〉

spring suspended type　　バネ懸架式

suspended arch　　吊りアーチ

suspension polymerization　　懸濁重合

suspension spring　　懸架バネ

suspension type triple action oil hydraulic press　　サスペンション形三駆動油圧プレス

(～で) できている，(～で) 構成されている

built-in : incorporated as part of a structure.	組込む，内蔵している

This control has **been built into** the machine.

この制御装置は，機械に**組込まれている．**

The tap-holding device has a clutch **built in** it.

タップ保持装置は，クラッチを**内蔵している．**

The oil hydrostatics are fed by an independent hydraulic system, and **built-in** safety devices provide an automatic output in the event of a pressure reduction.

油の静圧は，独立した油圧系統から送られている．圧力が低下した場合には，**組込まれている**安全装置が自動的に静圧出力を出す．

〈用　語　例〉

built-in antenna　　組込みアンテナ

build-in controller　　組込み形制御装置

compose : to form, to make up	形成する，合成する

Wheel teeth which **are composed of** two helical portions, one left-handed and the other right-handed, to avoid end thrust on the axle…….

車軸にスラストが作用しないように，2つのヘリカル部分（1つは左，他は右）で**構成されている**車輪歯車の歯は…….

The first titanium-based carbide cutting tools **were composed of** 42.5 percent titanium carbide, 42.5 percent molybdenum carbide, 14 percent nickel and 1.0 percent chromium.

最初の Ti 系超硬切削工具の**組成は**，炭化 Ti42.5%，炭化 Mo42.5%，Ni14%，Cr 1%で**あった．**

Typical simplified mechanical shaft seal theory assumes that the forces which maintain equilibrium between the seal faces **are composed of** spring, hydrodynamic, hydrostatic, and contact forces. Those tending to close the seal are the spring force plus a hydrostatic force due to stuffingbox pressure.

普通の単純化したメカニカル・シャフト・シールの理論では，つぎのように仮定する．シール面間の平衡を維持する力は，バネ力，流体力学的力，静水力学的力および接触力**からなる．**シールを閉じようとする力は，バネ力と，スタフィングボックス圧による静水力学的力の和である．

〈用　語　例〉

composition and resolution of forces　力の合成と分解

fiber composition in paper　紙の繊維組成

spectral composition　分光組成

---

comprise: to include, to consist of
to form, to make up

成り立っている，構成されている

The die head, a device that is clamped the lathe, **comprises** a cylindrical body containing chasers for cutting the thread.

ダイ・ヘッドは旋盤に固定される道具で，ネジ切削用チェーザを備えた円筒形基体で**できている．**

……. Because of their method of connection the former, which consists of a large number of turns of fine wire, is called a shunt or voltage winding, and the latter, which **comprises** fewer turns of heavier guage wire, is termed a series winding.

……．その接続の仕方によって，細い線を多数巻いたものでできている前者を分巻または電圧巻線と呼び，それより太いゲージ線で少ない巻数のもので**できている**後者を直列巻線という．

If the receiver consists of two units——one containing the receiver and power unit, the other **comprising** the loudspeaker——and when the controls operate through flexible cables, the set can sometimes be housed in the tool compartment under the bonnet.

もし受話器が２つのユニット——一方に受話器および電源ユニットが納まり，他方に拡声器が**入っている**——で構成され，かつ制御装置がフレキシブル・ケーブルを介して働く場合には，このセットをボンネットの下の工具用間切り内に収納できることがある．

N is an integrated NC package, **comprising** a milling machine and control system engineered to work together.

Nは１つの完全なNCセットで，一緒に働くように設計製作されたフライス盤と制御システムで**構成されている**．

India rubber solution——A semi-solid substance **comprising** mainly a solution of pure rubber in a solvent such as naphtha.

インドゴム溶液——主として，ナフサのような溶剤に純粋なゴムを溶かしたもの**でできている**半固体の物質．

The machine **is comprised** basically **of** a wheel lifter system, rotating spindle, an upper hold-down, and a measuring head assembly.

この機械の基本**構成要素**は，車輪吊上げシステム，回転軸，上部の持ち下げ装置および測定ヘッド・アッシ**である**．

Comprised of four separate modules, M-3 monitors all data to and from the data base.

別個のモジュール４つで**構成した**M-３は，データベースに出し入れするすべてのデータをモニターする．

……. This lack of engineering knowledge is a major reason why seal failures **comprise** the greatest portion of centrifugal pump maintenance costs.

……．この工学的知識の欠除こそ，シール損耗が遠心ポンプ保全コストの大部分を**占めている**主な理由である．

consist of : to be made up of

構成されている，（〜で）できている

Armature **consists** essentially **of** coils of insulated copper wire wounded over an iron centre.

電機子を**構成する**本質要素は，鉄心に巻付けた絶縁銅線コイルである。

Each machine **consists of** eight independently motorized drilling units.

機械はそれぞれ，8台の独立したモータ駆動の穴明け装置で**構成されている**。

These new seals **consist of** a metal case, an elastomeric element bonded to it, plus a nonwoven fabric shaft contact surface.

これらの新しいシールは，金属ケース，それに接着したエラストマ要素，および不織布製軸接触面で**できている**。

The system is a mechanical arm which **consists of** a robotic rotating arm with magnetic or vacuum grippers combined with a loading station to preposition the workpiece.

このシステムは，工作物をあらかじめ位置決めするためのローディング・ステーションを組合せた，磁気あるいはバキューム・グリップ付ロボット旋回腕で**構成した**メカニカル・アームである。

Abrasive wheel **consists of** individual grains of very hard material.

砥石は，非常に硬い材料の個々の砥粒で**成り立っている**。

The whole roll **consists of** one material. When the roll is cast, the molten material in direct contact with the cold chill will solidify first and the roll body surface will then **consist of** hard wear-resistant white iron.

ロール全体は，1つの材料で**できている**。ロールが鋳込まれるとき，冷いチルに直接接触する溶けた材料がまず凝固して，つぎにロール本体表面が硬い耐摩耗性の白鋳鉄を**形成する**。

Surfaces **consist of** numerous hill and

表面は，多数の山と谷で**形成**

valleys.

されている．

The work holding surface **consists of** a rotary table having radial T slots.

ワーク保持面は，放射状のT溝のあるロータリテーブルで**構成されている．**

Gear wheels are usually marked for retiming as shown. The marking may **consist of** punch dots, scribed lines, or etched marks on the gear teeth.

図示のように歯車は通常，ふたたびタイミングを合わせられるように，印をつける．この目印は，歯車の歯にポンチによる点，ケガキ線，あるいはエッチングした印**などである．**

The lifetime test **consists of** a continuous rotation at 24,000rpm at ambience temperature. At regular intervals, the deceleration time has been recorded. The lubricant loss has been determined by weight loss of the bearings.

寿命時間試験の**内容は**，室温で，24,000rpmの連続回転**である**．一定間隔で，減速時間を記録した．潤滑剤の減耗は，軸受の重量によって求めた．

The B provides rugged machining capabilities **consisting of** ¾″ drilling and tapping, one cubic inch milling metal removal in C 1040 steel, precise boring and accurate table positioning.

Bは，3/4インチ径の穴明けおよびタップ立て，C1040鋼を1 in^3フライス削り，精密中ぐりおよび正確なテーブル位置決めなどの強固な切削能力を**具備**している．

constitute : to make up, to form
: to be the components of
: to establish or be

構成する，(構成成分は)～である

On an average car the sparking plugs **constitute** only about one three-thousandth of the total weight, yet if their dimensions are only very slightly unsuited to the particular engine or if they develop

普通の自動車では，スパークプラグは全重量の約1/3,000を**占める**に過ぎないが，もしその寸法がエンジンに対しわずかでも不適当であったり，あるいは

a quite invisible fault the whole car may be put out of action.

まったく目に見えないような欠陥があると，車全体が動かなくなることがある．

Usually in practice Ze **constitutes** 90 percent and Za **constitutes** 10 percent of the total metal-removal rate Z.

通常，実際にはZeが全切削量Zの90%，Zaが10%**である．**

〈用　語　例〉

three-dimensional constitution's polymer　3次元構造重合体

contain : to have within itself
: to consist of

もっている，中にある

An important area of instrumentation **contains** delicate components (electronic devices, precision gyro bearings, etc.) which are extremely sensitive to atmosphere-borne particulate contamination.

計装の中核部には，大気中の微粒子汚濁にきわめて敏感な繊細な構成要素（電子装置，精密ジャイロ軸受など）が**ある．**

The control station **contains** all the controls required for use of machine operation.

制御ステーションには，機械の操作に必要な制御装置すべてが，**納められている．**

Adjusting cap assembly **contains** adjusting screw and locknut for altering length of piston travel.

調整キャップ機構には，ピストンの動きの長さを変えられるように，調整ネジおよび固定ナットが**付いている．**

The control **contains** an electronic probetype sensing element that is mounted so it contacts a reference surface on the workpiece while the boring tool is being fed.

制御装置には，中ぐりバイトが送られている間中，工作物の基準表面に接触しているように取付けた電子プローブ形検出素子**が含まれている．**

D toolpost grinder kits **contain** toolpost grinder and accessories capable of handl-

D刃物台研削キットは，刃物台研削装置と，内研，外研両方

ing a wide range of both internal and external grinding applications.

を広範囲に処理できる付属品で**構成されている.**

Milling machines are of the horizontal or the vertical type. A commonly employed horizontal machine is the knee type. It comprises a massive column which **contains** the gearbox and spindle-drive motor and is provided with bearings for the spindle.

フライス盤には，横形または立形がある．普通使用される横フライス盤はニータイプである．これは，歯車箱および主軸駆動モータを**もった**，がっしりしたコラムでできていて，主軸用軸受も付いている．

The turret head **contains** six 5/8″-16 male threaded spindles.

タレット・ヘッド**の中**には，6本の5/8″-16の雄ネジを切ったスピンドルが**ある.**

The cylinder **contains** a high force spring actuated plunger.

シリンダの中には，強力バネ作動のプランジャが**納まっている.**

Common solder usually **contains** one part of lead to one part of tin, though different solders exist which are composed of lead alloyed with tin, the proportions of the latter varying from 30 to 70 percent.

普通のハンダの**成分は**，通常錫1に対し鉛1**である.** これと違ったハンダもあり，錫の割合が30%から70%までといろいろな鉛との合金である．

All of the software you need for any type of programming **is contained** on standard, magnetic-tape cassettes.

どんな種類のプログラミングに必要なソフトウェアすべてが，標準の磁気テープカセットに**納められている.**

equip: to supply with what is needed. **装備している，付いている**

The U **are equipped with** separate motor drive for spindle, for drive shaft, and for individual attachments.

Uは主軸用，駆動軸用および個々のアタッチメント用に別々のモータ駆動装置を**備えている.**

The M-10 **is equipped with** a ten tool sequential automatic tool changer, it features, exceptionally high speed with a chip time of less than 5 seconds.

M-10には，工具10本用のシーケンス制御の自動工具交換装置を**装備している**．実切削時間が5秒以下と，稀にみる高速がその特徴である．

B Automatic Hydraulic Machines **are equipped with** a 4-speed AC motor and a 2-speed electric clutch which provides 8-speeds per spindle.

B自動油圧機械は，4速の交流モータと2速の電気式クラッチが**付いていて**，これで主軸はいずれも8速になる．

A heavy duty mill, the Model V-2, **is equipped with** one-piece column and base, coolant system built into the base, work light, hardened and ground gears and shafts, and 1.5～2-hp two-speed motor.

重切削用フライス盤V-2形には，一体形のコラムとベース，ベース組込み形冷却液システム，工作物照明灯，焼入れ研削の歯車と軸，および1.5～2HPの2速モータが**付いている**．

Along with the pallet changer, the 20″ cube machining center can **be equipped with** a robot, workpiece storage towers, and pallet conveyor systems for unmanned production runs of workpieces requiring very precise machining.

パレット・チェンジャとともに，この20インチ立方のマシニングセンタには，きわめて精密な機械加工を要求される工作物の無人連続生産用に，ロボット，ワーク保管塔およびパレット・コンベヤシステムを**装備する**ことができる．

……. It can be done on a lathe with the aid of a thread-rolling head, which **is equipped with** three ribbed rollers that can move in and out radially. The rollers open out automatically at the end of the operation.

……．それは，半径方向に出入り可能なリブ付ローラ3個を**備えた**ネジ転造ヘッドを利用して，旋盤により行なえる．このローラは，作業が終わると自動的に開く．

A vertical milling machine **equipped with** variable feed and speed drives was

円周端のフライス削り試験を実施するために，送りおよび速

used for performing the peripheral end milling tests. The milling machine **was** also **equipped with** a tachometer for observing spindle speed.

度可変駆動装置の**付いた**立フライス盤を使用した．この立フライス盤はまた，主軸速度を見るためのタコメータも**備えている．**

Equipped with specially selected work-head bearings, the machine grinds the center bore.

特に精選した主軸台軸受を**用いた**この機械で，センタ穴を研削する．

feature : one of the named parts of the face (e.g. mouth, nose, eyes) which together make up its appearance, a distinctive or noticeable quality or a thing
: to be a feature of or in

特徴ある構成要素，特徴である，～がある

The M-7 manufacturing system **features** a robotic arm which can move in four directions with a payload of up to 55lbs.

M-7製造システムには，55ポンドまでの搭載荷重で，4方向に動くロボット・アーム**がある．**

The D micrometer No.2 **features** a liquidcrystal display and provides inch or metric measurements. The scale is 0 to 1″ or 0 to 25mm in 0.0001″ or 0.001mm graduations.

Dマイクロメータ No.2には，液晶表示装置**があり**，インチまたはミリ測定ができる．目盛は0.0001インチ目盛りで範囲0～1インチ，または0.001mm目盛りで範囲0～25mmである．

The B-8 single-spindle machining center has been designed for the toolroom and precision job shops and **features** all cast iron construction, a 3-point bed mounting, gearless AC spindle with water-cooled motor, an 18-tool changer, and CNC control.

B-8単軸マシニングセンタは，工具室および精密加工工場用に設計したもので，**主な特徴は**全鋳鉄構造，3点支持ベッド，歯車なしの水冷交流モータ付主軸，18ツールのツール・チェンジャおよびCNC制御**である．**

The S-10 Machining Center **features** an integral console conveniently located so that the operator has easy access to both the controls and the automatic tool changer. The console includes all the necessary controls for either automatic or manual operation.

このＳ-10マシニングセンタには，作業者が制御装置，自動ツール・チェンジャの両方に簡単に近づけるように置いた多機能コンソールが**ある**．このコンソールには，自動・手動操作のいずれかに必要な制御装置すべてが入っている．

G-8 series tool grinders from Y are available with NT40 or NT50 spindle tapers, and **feature** a cast-iron base, bellows-protected moving parts, and angular contact bearings for the spindle head.

Ｙ社製のＧ-８シリーズ工具研削盤には，主軸テーパＮＴ40またはＮＴ50のものがあり，**主な**部品は，鋳鉄のベース，ベローズで保護した可動部品，主軸台用アンギュラ軸受など**である**．

Vertical Mill. The L-3 for CNC operations **features** a 54×11″ table with 28″ longitudinal capacity, 13¼″ cross travel. Includes B control three-axis simultaneous contouring system.

立フライス盤．CNC 作業用L-3には，左右の動き28インチ，前後の動き13¼インチのテーブル（54×11インチ）**が付いている**．Ｂ制御の３軸同時輪郭削りシステムも備えている．

A 1,000-ton capacity press has been added to the Series 27 line of drawing and forming presses. Equipped with a 250-ton hydraulically-operated cushion, the press has a slide stroke of 30″, cushion stroke of 12″ and **features** remote digital set-point stroke controls and digital readout of stroke position. Ram speeds per minute include a rapid advance of 600″; pressing of 4″ to 37″; …….

引き抜き・成形プレスの27シリーズに，1,000トンプレスが加えられた．250トン油圧作動のクッションの付いたこのプレスは，スライド行程30インチ，クッション行程12インチで，遠隔操作ディジタル式設定点長さ制御装置，およびディジタル読取り装置を**備えている**．ラムの毎分速度は，急速前進600インチ，プレス作業４～37インチ，……などである．

The Y knee-type CNC vertical milling machine **features** a spindle head equipped with a ball screw Z-axis drive for heavy cuts and repeatability.

Yニータイプ CNC 立フライス盤で**特に目につくのは**，重切削および高繰返し精度を実現するための，ボールネジ Z 軸駆動装置の付いた主軸頭**である**.

The L-7 **features** a 10-hp (7.5-kW) drive motor, a D1-8 camlock spindle nose, and a 3 1/6″ (80.4-mm) hole through the spindle. The lathe is equipped with an electromagnetic clutch, a built-in coolant system, and more.

L－7の**特徴は**，10馬力駆動モータ，カム固定式主軸端および主軸の3⅙インチ通し穴**である**．この旋盤は，電磁クラッチ，組込み形の冷却システムなどを備えている．

The CNC Duplex horizontal twin mill **features** 3 axis CNC control for head and cutter positioning utilizing DC servomotors and ballscrew drives. Workpiece holding devices, composed of a special air indexer and tailstock assembly, rotate the steel part upon command from the CNC pendant to perform the cutting cycle in less than one second.

この CNC 複式 2 軸横フライス盤は，直流サーボモータおよびボールネジ駆動装置を使って，ヘッドおよびフライスを位置決めする 3 軸 CNC 制御装置を**備えている**．特殊なエア式割出し装置と心押し台機構とで構成したワーク保持装置は，切削サイクルを 1 秒以下で行なうように，CNC ペンダント（吊下げ式制御盤）からの命令があると，鋼部品を回転させる．

The collet-type workholding fixture **features** a 360° graduated dial, an index locking pin and 24-division index plate, to ±10 seconds accuracy, and adjustable stops and locking pin for locations between indexes.

このコレット形ワーク保持用取付け具の**主要部品は**，360°目盛りのダイアル，割出し固定ピン，24分割の割出し板で精度は±10秒，割出ししない間位置決めしておくための調整式ストッパと固定ピン**である**．

Available for quick delivery, this precision, highly reliable and productive 2

即納品の精密，高信頼性，高生産性 2 軸 CNC 旋削盤は，つ

-axis CNC turning machine **features**:
- 6-1/4″ chuck
- 1-3/8″ collet capacity
- 4,800rpm
- ……

ぎのような**特徴を備えている．**
- チャック　6¼インチ
- コレット容量　1⅜インチ
- 4,800rpm
- ……

PRECISION CYLINDRICAL GRINDER Model RMG 4

Features
- Swing 4″
- 10″ between centers
- ……

精密円筒研削盤 RMG 4 形の**主要諸元は，**
- 振り　4インチ
- 芯間距離　10インチ
- ……

The L-15 lathe has a 12-station turret, 15hp drive motor, ……. Standard **features** include a 15¾″ swing, slant bed construction, a spindle with infinitely variable speeds from 70 to 3,800rpm, and a 2 1/5″ bore for big bar work.

L-15旋盤には，12ステーションのタレット，15馬力駆動モータ，……などを備えている．標準の**主な仕様は，**振り15¾インチ，傾斜ベッド構造，70～3,800 rpm 無段変速の主軸，および大きな棒材作業用の2⅕インチの穴径で**ある．**

Outstanding **features** of the K include : · 2 HP vertical drive, · 3 HP horizontal drive, · Builtin coolant system, · ……, · Net weight 3,000lbs.

Kの優れた**特徴は，**・2HP 立駆動装置，・3HP 横駆動装置，・組込み冷却剤系統，・……，・正味重量3,000lb，**などである．**

…… and safety **features** include a foot-operated spindle brake, emergency stop button, motor safety switch, and a shear pin to protect against overloads.

……主な安全装置**には，**足踏み式主軸ブレーキ，緊急停止ボタン，モータ安全スイッチ，およびオーバーロード防護シャーピンなど**がある．**

Cutting-off lathe **features**:
- Setup——30 to 45 minutes

突切り旋盤の**特徴**：
- 段取り——30～45分

・Choice of Tooling——High speed steel, roller or new reversible throw-away carbide	・ツーリングの選択——高速度鋼，ローラまたはリバーシブル・スローアウェイ超硬合金．
・……	・……

Standard **Features**:
・Automatic lubrication system
・……
Special **Features**:
・Easy Speed and feed change
・Automatic lube of apron and ways
・……

標準仕様：
・自動潤滑系
・……
特色のある**仕様**：
・速度および送りの変更容易
・エプロンおよび案内面の自動給油．
・……

form: to mould, to produce or construct : to bring into existence, to constitute	**形づくる**

NITROGEN.——One of the gaseous constituents of the atmosphere **forming** about four-fifths of the total volume of the atmosphere.

窒素——大気の全容積の約4/5を**形成する**気体構成成分の1つ．

……. This **forms** an electric circuit, stopping the machine or giving an alarm.

……．これで，機械を停めるかあるいは警報を出す，電気回路が**できる**．

furnish: to equip (a room or house etc.) with furniture : to provide or supply	**備え付ける**

The mounting **is furnished with** an additional sliding seal consisting of a bronze ring.

この架台には，さらに青銅リングでできた滑りシールを**設けてある**．

For such use it **is furnished with** a

そのような用途向けには，回

rotary index head and …….

転割出しヘッドを**装備し**，…….

> **have**: to be in possession of (a thing or quality), to possess, to contain.
>
> **もっている，～がある**

The L-5 75 hp slant bed CNC lathe **has** two turrets, with 20-tool capacity, four axes, a massive slant bed and CNC controls.

L－5，75馬力，傾斜ベッドCNC旋盤は，ツール容量20本，4軸の2つのタレット，がっしりした傾斜ベッドおよびCNC制御装置で**構成されている**．

It **has** a toolholder with a compound graduated in 2-degree increments for direct and accurate setting of grinding angles.

それには，研削角度を直接かつ正確にセットできるように，2度単位で目盛った複式刃物台付きのバイトホルダが**付いている**．

A horizontal milling machine is known as the manufacturing type which is characterized by **having** a work table that is fixed in height, the spindle being vertically adjustable, since it is mounted in a head that can be moved up or down the column of the machine.

横フライス盤は，いわゆる生産形で，その特徴は，高さが固定のワークテーブル，立方向調整式の主軸（機械のコラムを上下に動かすことができるヘッドに取付けられているから）から**なっている**ことである．

A-25 system **has** three components: the H calculator, a plotter, and a tape punch. In addition a large 15K memory and various ROMs are employed.

A－25システムは3つの構成要素から**なっている**：H計算機，プロッタおよびテープ穿孔装置．加えて，15kの大容量記憶装置および各種のROMを採用している．

The normal type of carburettor, which **has** jets of fixed size, **has** three principal parts. These are for slow-running, normal

固定寸法のジェットが**付いている**普通の気化器は，3つの主要部品から**なっている**．すなわ

highspeed running, and for changing over from one of these conditions to the other.

ち，低速走行用，定常の高速走行用，およびこれらの状態の1つから他の状態への変換用である．

include : to have or regard or treat as part of a whole
: to put into a certain category or list etc.

（～の中に）ある，～もある，付いている

The wheelhead **includes** a 4hp, two speed motor that drives the 8″ grinding wheel by means of a Polyvee belt.

ホイールヘッド（砥石頭）には，ポリヴィ・ベルトで8インチ砥石を駆動する4馬力，2速モータが**付いている**．

The S line of compact power workholding components **includes** over 70 models and sizes that can operate comfortably at up to 7,500 psi fluid pressure.

コンパクトな動力作動工作物保持部品のS系列**には**，7,500 psiまでの液圧で快適に操作できる，70種以上の形式および寸法が**ある**．

The line consists of an induction heater **including** a hot shear unit, a press line with the biggest die-forging press in the world, press power 16,000 tonnes, and a cooling line.

このラインは，熱間剪断装置**が含まれている**高周波加熱装置，世界最大の型鍛造プレス（プレス力16,000トン）のあるプレスライン，および冷却ラインで構成されている．

The total system **includes** a lightweight overhead conveyer and a S sorting system.

システム全体には，軽量のオーバーヘッド・コンベヤとS選別システムが**含まれている**．

Two video display terminals and a dotmatrix printer **are included in** the system.

このシステム**には**2つのビデオ表示装置および1つのドット・マトリックス印字装置**がある**．

Components **include** low height edge clamps and self-adjusting work supports that advance and lock automatically.

構成部品**として**，高さの低いエッジ・クランプ，および自動的に進みかつ固定する自己調整形の工作物支持具**がある．**

Accessories **include** a 5″ diameter magnetic chuck, a 5¼″ face plate, and a 3-jaw chuck with reversible jaws.

付属品の**内訳は**，直径5インチの磁気チャック，5¼インチの面板および三ツ爪チャック（可逆爪付き）**である．**

Features **include** an automatic toolchanger, horizontal/vertical swing toolhead and a rotary table.

主な装置**には**，自動工具交換装置，水平・垂直旋回のツールヘッドおよびロータリ・テーブル**がある．**

Such standard features as induction hardened headstock gears, precision spindle bearings, ……, and induction hardened and ground bedways **are included in** the lathe.

この旋盤**には**，高周波焼入れ主軸台歯車，精密主軸軸受，……および高周波焼入・研削ベッド案内面などの主要な標準部品**がある．**

Fig. 5 ——Groove dimensions that are most critical in determining cutting performance **include** land width, groove width, groove angle, and groove depth.

図5——切削性能を決める上で最も重要な溝寸法諸元**には**，ランドの幅，溝幅，溝角，および溝深さ**がある．**

Iron-base superalloys **include** various types of martensitic, semiaustenitic, and austenitic steels. Proprietary materials in this classification **include** Discaloy, Incoloy 800, 19-9 DL, A286, and N-155.

鉄系スーパーアロイ**には**，マルテンサイト，セミオーステナイト，オーステナイト鋼などいろいろな種類**がある．**この分類に入る専売特許の材料**には**，Discaloy，Incoloy 800，19-9 DL，A-286，およびN-155など**がある．**

Machine specifications **include** a 10″

機械の仕様は，チャック寸法

chuck size, 24″ shaft length, 2″ bar capacity, maximum swing of 22¾″, cross travel of 6.5″ and longitudial travel of 24.4″…….

10インチ，シャフト長さ24インチ，バー能力2インチ，最大振り22¾インチ，前後の動き6.5インチ，左右の動き24.4インチ……**などである**．

The grinder is capable of grinding cutting tools without special accessories. Grinding capabilities **include** ball, taper, ……, and high speed and carbide end mills ; reamers and tool bits.

この研削盤は，特殊な付属品を使うことなく切削工具を研削できる．研削できるのは，球，テーパ，……および高速および超硬のエンド・ミル；……リーマおよび完成バイト**などである**．

The X-axis table travel of the E-4 is 500 mm as is the Z-axis table travel. Travel rates **include** a rapid traverse of 10m/min and a feedrate that varies from 1 to 2,997 mm/min.

E-4のテーブルX軸方向の動きはZ軸方向のテーブルの動きと同じ500mm．動きの速さには，早送り10m/min，および1mm/min〜2,997mm/minの範囲で可変の送り速度**がある**．

Options **include** internal grinding attachments, static wheel balancing stand and 3-jaw chuck.

オプション(選択できる品目)**として**，内面研削アタッチメント，静砥石バランシング台，および三ツ爪チャック**がある**．

Optional equipment **includes** a collet storage rack, bar feeders, Brown & Sharpe No. 21 spindle headstock, magazine parts loader, part take-off chute and a vertical cut-off slide.

オプションの装置**には**，コレット保管ラック，バー・フィーダ，B&S No. 21主軸台，マガジン・パーツローダ，部品取出しシュートおよび立切断スライド**がある**．

Options **including** conveyors and handling equipment are available.

コンベヤおよびハンドリング装置**など**のオプションは在庫**がある**．

Heavy ribbed knee and saddle construction for rigidity **includes** hard chrome plating on X-Y ways and gibs.

剛性をもたせるために強力なリブを付けたニーおよびサドル構造には，硬質クロムメッキしたX-Y案内面およびジブも**含まれている**.

Applications **include** bar work, chucking, and second operation drilling and milling.

用途は，棒材作業，チャッキング，および2次作業穴明けおよびフライス削りなど**である**.

There are many forms of data collection **including** : strip and circular chart recording, punched tape and digital printers of either the tape or carriage type.

データ収集**には**，つぎのような多くの方式**がある**：帯状および円形チャート記録，穿孔テープ，およびテープ式あるいはキャリッジ式いずれかのディジタル・プリンタ.

incorporate : to take in or include as a part or parts of itself.

～もある，取り入れる，組み入れる

Assembly operations. The final assembly machine **incorporates** 12 active stations, plus four open stations. These open stations are to accomodate possible future changes in manufacturing operations.

組立作業．最終組立機**には**，12の実働ステーションと4つの空きステーション**がある**．この空きステーションは，今後ありうる製造作業の変更に適応するためである.

The gauge **incorporates** a micro-indicator.

このゲージ**には**，マイクロ・インジケータ**も付いている**.

The spindle head **incorporates** constant mesh helical gears.

このスピンドル・ヘッドには，常時嚙合いのはすば歯車**もある**.

Where "Log Out" is required for permanent records a printer type recorder **is incorporated into** a J Control Center.

永久記録保存のために，計算機操作の記録が必要な場合には，J制御センターにプリンタ方式の記録装置を**組み入れる**.

……. The device **incorporated** for this purpose is the universal joints, of which there are many types.

……. この目的で**組み入れられる**装置が自在継手で，多くの種類がある.

The pallets **incorporated** in this line are somewhat unusual due to the auxiliary fixtures mounted on them. Each is equipped with part nests for carrying two gears, a transfer shaft, and a slave tool.

このラインに**組み入れられた**パレットは，それに補助取付け具が固定されているので，若干普通のものと異なる．各パレットは，2つの歯車，伝導軸，従属工具を載せるための部品巣箱を備えている.

Input sheets are sent on a monthly base to the computer center for **incorporation** into the master catalog.

入力シートはマスター・カタログに**組み入れる**ため，月単位で計算センターに送る.

To accommodate billets up to 190mm square section, and rolling billets up to 2, 200mm long, the forging roll unit **incorporates** horizontal operation with a roll diameter of 1,000mm.

断面190mm角までのビレット，および長さ2,200mmまでの圧延ビレットに適応できるように，この鍛造圧延装置はロール直径が1,000mmで，水平方向操作を**採用している**.

involve: to contain within itself
: to include or affect in its position

含む，～もある，伴う

The most drastic example of the contribution to production by synthetic lubricants **involves** some 6,000-horsepower vertical electric motors driving the reactor-cooling water pumps in a nuclear power plant.

合成潤滑剤が生産に寄与した最も顕著な事例に，原子力発電所の原子炉冷却水ポンプを駆動する6,000馬力立形モータ**がある**.

A typical part turned on E **involves** feeding bar stock through the spindle to a specified length, rough facing and rough OD turning, drilling, rough boring to within 0.002″ of depth and ID, and finish boring to specified depth and ID. Stock hardness ranges from 220 to 240 Brinnell.

Eで旋削する典型的な部品では，棒材を主軸を通して指定の長さにまで送り，荒面削りおよび荒内径旋削，穴明け，深さと内径が0.002インチまでの荒中ぐり，および指定の深さおよび内径への仕上げ中ぐりなどの作業が**ある**．

This is not a cutting method and **involves** no removal of metal, the thread being formed by plastic deformation. Not only is there a saving in metal, but the rolling operation causes cold working and thus improves the mechanical properties of the thread.

これは，切削する方式ではないため，金属を除去すること**なく**，ネジを塑性変形によって成形する．金属の節約だけでなく，転造作業によって加工硬化することによりネジの機械的性質が向上する．

The machines were continued in production for eight hours and no downtime **was involved**.

この機械は，8時間生産を続けたが，その間まったく故障停止時間が**なかった**．

……. Communication systems enable the operating engineer to talk with personnel at the actual equipment or system location **involved**.

…….操作技術者が，装置あるいはシステムの現場で作業に**従事している**人と通信システムで話をすることができる．

make up: to form or constitute, to put together.

つくる，形成する，構成する

The photograph in Fig. 1 shows the two units, machine and control cabinet, that **make up** the tapping torque test machine.

図1の写真は，タップ立てトルク試験機を**構成する**2つのユニット，すなわち，機械と制御キャビネットである．

P builds a line of 20″ geared-head drill presses, some of which can be mounted on

P社は，20インチ歯車式ヘッドのボール盤シリーズを製作し

production tables to **make up** a multispindle unit (they can be used individually).

ている．その中には，多軸ユニットを**形成する**ように，作業用テーブルに取付けできるものもある（別々に使うこともできる）．

You can't **make it up** in the laboratory by putting all the proper chemicals together.

それは，適切な薬品をすべて一緒にしても，研究室で**作る**ことはできない．

A plug **is made up of** three main parts, the central electrode, the insulator and the body.

プラグは，３つの主要部品，つまり中央電極，絶縁物およびボディで**できている**．

Another answer to multiple hole drilling is the multiple spindle drill press (also referred to as a gang drilling machine). These rigs **are made up of** a number of individual drill presses mounted on one table.

多数の穴をあけるもう１つの解決策は，多頭ボール盤（また，ギャングボール盤ともいう）である．これらの装置は，１つのテーブルの上に何台かのボール盤を個々に取付けて，**作り上げる**．

Those registers which can be controlled by the operator **are made up of** tiny units consisting of a capacitor and a switching transistor which connects it to the other circuitry.

オペレータが制御できるこれらのレジスタは，小さいユニットで**できており**，コンデンサと，他の回路とに連結できるスイッチング・トランジスタで構成されている．

MoS_2 has a lamilar structure, each lamina is a sandwich **made up of** two layers of surfur atoms with a layer of Mo atoms between.

MoS_2は積層構造である．各薄層は，Ｓ原子の２層とその間のMo原子の１層で**形成された**サンドイッチ構造である．

On a typical five stand mill, **make-up** water is required at a rate of about 800 to

典型的な５スタンド圧延機では，**補給**水が毎時800～1,000ガ

1,000 gallons per hour. ロン必要である．

possess: to hold belonging to oneself, to have or own	もっている

An insulated conductor **possesses** an external protective covering of metal.

絶縁した導線には，外側に金属の防護覆が**付いている**．

provide: to cause, a person, to have possesion or use of something, to supply, to make preparation for something	設けてある，付いている，用意されている

An exceptionally large capacity coolant reservoir **is provided** in the base of the machine.

格別大容量の冷却液タンクが，機械のベースの中に**設けてある**．

The machine <u>provides</u> an additional 19-1/2″ between centers. Also **provided** as standard **are** a program-controlled tail-stock body and quill. Special steady rests are optionally available.

この機械は，心間距離をさらに19½インチ大きくすることが<u>できる</u>．また，標準品として，プログラム制御の心押し台本体およびクイルが**付いている**．特殊振れ止めは，オプションで在庫がある．

These chucks **are provided with** a taper shank for locating in the drill-press spindle.

これらのチャックには，ボール盤主軸に位置決めするためのテーパ・シャンクが**付いている**．

The back of the dovetail ways on the saddle and spindle head **are provided with** hardened and ground steel jibways for minimum wear and long life.

サドルおよび主軸頭のアリ溝案内面の背面には，摩耗が少なく寿命が長くなるように，焼入れ研削した鋼のジブウェイを**備えている**．

……. It comprises a massive column

……．これは，歯車箱と主軸

which contains the gearbox and spindle-drive motor and **is provided with** bearings for the spindle.

駆動モータを収めたがっしりしたコラムでできていて，主軸用の軸受も**付いている．**

The machining head has a total vertical travel of 25mm and **is provided with** a pneumatically operated static loading system, which provides precise control.

この加工ヘッドは，縦方向の動き25mmで，空気作動の静負荷システムを**備え**ており，これで精密な制御ができる．

The panel **provides** START and STOP button for the spindle and START, STOP and JOG button for the drive shaft.

パネルには，主軸の始動および停止用ボタン（スイッチ）ならびに駆動軸の始動，停止およびジョグ用ボタンが**ある．**

Speed change gears, at lower left side of machine, **provide** for changes of spindle speed in either direction.

主軸速度をどちらの方向にも変えられるよう変速歯車装置が，機械の左下側に**付いている．**

A selector switch **is provided** for each of the six spindles.

6つの軸おのおのに，セレクタ・スイッチが**付いている．**

Alarms **are provided** for a system trouble indication as well as power failure.

電源故障用およびシステムの故障表示用に，警報装置が**付いている．**

Controls **are provided** to maintain oil and air temperatures at proper levels and, in some cases, to maintain correct oil level.

油および空気温度を適切な水準に維持するよう，また場合によっては，正しい油面を維持するように，制御装置を**備えている．**

Readouts on the panel **provides** digital displays of the momentary values of the radial and lateral runouts on the inner and outer bead seats. Analog meters also **are provided** to determine the TIR values of

パネルの読出装置は，内側および外側のビード（溶着金属）の座の半径方向および横方向の振れの瞬間値をディジタル表示**する．**また，アナログ計がこの

these same readings.

同じ読みのTIR値（インジケータの全振れ）が求められるように**付いている．**

supply : to give or provide with (something needed or useful)
: to make up for, to make available for use

供給する，設ける

In case the pump **has been supplied** with a geared motor the grease in the gear unit should be changed after about 8,000 hours of operation and not later than after 2 or 3 years of operation.

ポンプが，ギアードモータ**付きの**場合には，歯車装置のグリースは，運転約8,000時間および2年または3年以内に交換すること．

This **is supplied**, together with the upper bearing, as a unit with a small axial clearance, accurately adjusted by means of an intermediate spacer sleeve of caliculated length.

これは，上方の軸受と一緒に，長さ補正した中間スペーサ・スリーブによって正確に調整した，小さい軸方向隙間のユニットとして**使うようにできている．**

with : having, characterized by

付いた，持った

High productivity CNC Machining Center——**with** H microprocessor control and 24-tool automatic tool changer.

高生産性CNCマシニングセンター——Hマイクロプロセッサ制御装置および24本の自動工具交換装置**付き．**

The advantages of ball bearing **with** seal permanently built-in have become recognized.

シールを恒久的に**組込んだ**玉軸受の利点が，認められるようになってきた．

M also sells spiral and spade-type drills made from high-speed steel **with** 9.5% cobalt, solid carbide, …….

M社はまた，Co 9.5%**含有の**高速度鋼，ムクの超硬……でつくったスパイラルドリルおよびスペードドリルを販売している．

できる，できない，してもよい（可能，能力，余裕）

ability : the quality that makes an action or process possible, capacity or power to do something　　できる，能力，資質

The user **has** the **ability** to edit programs at the machine and **may** choose manual / single block / automatic operation.

ユーザーは，機械でプログラムを編集**でき**，また手動／シングル・ブロック／自動操作のどれかを選ぶことが**できる**．

The microcomputer provides the **ability** to measure and analyze, as well as the **capacity** to receive and store operator instructions.

このマイクロコンピュータは，測定し解析する**能力**，ならびにオペレータの命令を受けて，記憶する**能力**をもっている．

able : having the ability to do something　　できる

With this information the designer will **be** better **able to** design the machine tool so that the natural frequency …….

この資料を使って，設計者は，固有振動数が……であるように工作機械をよりよく設計することが**できる**．

Grade 9 ceramic-coated tungsten carbide inserts combine the features of ceramic and carbide in one insert capable of cutting at high speeds while **being able to** withstand shock.

9級セラミック・コーティング超硬合金のインサートは，衝撃にも耐えることが**できる**一方で，高速切削も可能なインサート工具で，セラミックと超硬合金の特徴を合わせもっている．

allow : to permit, to tolerate　　できる，してもよい

The infinitely variable cutting speed of from 52 to 280 fpm is adjustable by remote

52～280fpm 無段変速の切削速度は，制御盤から遠隔制御で

control from control panel and **allows** cutting of most materials with high or low tensile strength.

調整できるので，高あるいは低引張り強さまでのほとんどの材料を切削**できる**.

Dial-in elecronic control to **allow** fast, easy setups.

段取りを早く楽に**できる**ダイヤル・イン電子制御.

Other features include clutches on the X and Y axis drives to **allow** full manual milling capabilities, a power feed mode for straight milling cuts and a pressure control manifold which **allows** the use of single pressure power unit in this three pressure system. The system **can** be adapted to many different types of mills from small, toolroom mills to large airframe profilers.

そのほか，つぎのような特徴がある．手動フライス削り能力がフルに発揮**できる**ようにしたXおよびY軸駆動装置，直線フライス切削用の動力送り方式と3圧力系統に1つの圧力出力装置が使**える**圧力制御マニホールド．この方式は小は工具室のフライス盤から大は航空機機体のならい盤まで，多様なフライス盤に適合させることが**できる**.

This technique **allows** diagnostic execution during normal machine operation **enabling** production to continue while fault testing.

この技法は，機械加工を普通にしている間に診断をすることが**でき**，故障を調べている間も生産を続けることが**できる**.

This system **allows** high speed data transmission unaffected by electrical fluctuations or noise.

このシステムは，電気的変動や雑音に影響されることなく，高速データ伝送が**できる**.

There is an RS-232-C interface that **allows** attachment of both input and output devices.

入力および出力装置の両方を取付けることの**できる**インターフェイス RS-232-C がある.

A 12-station crown turret above the spindle axis **allows** the mounting of very long boring bars and a 6-station turret below the spindle axis **permits** use of

主軸より上方にある12ステーションクラウン・タレットは，きわめて長い中ぐり棒を取付けることが**でき**，主軸より下にあ

additional tools for external machining.

る6ステーション・タレットは，外面切削用のツールを追加して使うことが**できる.**

The unit **allows** a printing speed of 30 characters per second as compared to the 10 or 15 cps typically available.

この装置は，普通のものが毎秒10あるいは15文字であるのに対し，30字の速度で印刷**できる.**

Now if the nose **is allowed** to rest on the side of the nut or some other part of the car, and pressure is applied to the handles, great leverage can be applied.

先端をナット側面か車の何か他の部分に静置することが**でき，**圧力をハンドルに加えれば，大きなこじり力がかけられる.

The Power Lift **allows** one man to load and position workpieces, tools, fixtures, inserts, dies, and jigs up to 250lb.

このパワーリフトは，1人で，250ポンドまでワーク，バイト，取付け具，インサート，ダイス，治具を載せ，かつ位置決めすることが**できる.**

Conversational language **allows** even a beginner to complete machining center programs in less than 5% of the time required for conventional CNC equipment.

会話形言語だから，初心者でも，普通CNC装置の場合の5%以下の時間で，マシニングセンタのプログラムを完成することが**できる.**

The Nibbler rotary broach attachment **allows** many different shapes to be broached as part of many screw machine operations.

この回転ブローチ・アタッチメントによって，多くのネジ切り作業の一部として多種類の形状をブローチ加工することが**できる.**

If carefully carried out this procedure should **allow** the bearings to be pressed out of their housings.

注意すれば，この要領で，軸受をそのハウジングから押出すことが**できる.**

Splined shaft is a shaft in which keyways

スプライン軸とは，(それに)

or slots are cut, thereby **allowing** a gear-wheel to be driven by the shaft and at the same time to be moved along the shaft at will.

キー溝または溝を切った軸で，これによって，歯車を駆動すると同時に，軸に沿って思うままに動かすことが**できる**．

Levers on the front of the headstock **allow** various rotational speeds to be selected.

主軸台前面のレバーによって，いろいろな回転速度を選ぶことが**できる**．

In this way **it is possible to** obtain good manoeuvrability with the additional advantage that the propeller **can** be rotated in one direction, at constant speed. This in turn **allows** a low-cost synchronous electric motor to be used as the power source.

このようにして，プロペラが一方向，定速で回転**できる**という利点が加わり，高い操船性を得る**ことができる**．これによってさらに，動力源として低コストの同期モータを使うことが**できる**．

Ball assembly **allows** the index table to float a limited amount and **permits** index plungers to freely and accurately find their locations in bushings in the J table.

ボール・アセンブリによって，割出しテーブルはある限られた量浮くことが**でき**，割出しプランジャは自由かつ正確に，テーブルのブッシュに位置決めすることが**できる**．

Mode switches which are used to select the desired cycle of operation, **make** additional flexibility **possible** by **allowing** the operating engineer to select any one of up to eight standard modes of operation.

所望する作業サイクルを選定するために使うモード・スイッチによって，操作員は８種の標準作業モードの１つを選定**できる**ようになるため，よりフレキシビリティを増すことが**できる**．

In this case the dynamo is mounted eccentrically, so as to **allow** of adjusting the tension, or it is mounted on a sliding spigot for the same purpose.

この場合，ダイナモは張力の調整が**できる**ように偏心して取付ける．あるいは同じ目的で，摺動するスピゴットにダイナモを取付ける．

Keyway and Tee slot locations **allow for** easy table tooling using simple clamps or vises.

キー溝およびT溝の位置は，簡単なクランプまたはバイスを使って，容易にテーブル上のツーリングができるように**配慮されている**.

The overall design of the table **allows for** operator convenience in loading and unloading.

テーブル全体の設計は，作業者がローディング/アンローディングを楽にできるように**してある**.

A clearance is always left between the tappet and the valve or push-rod to **permit** of the valve closing properly and to **allow for** slight expansion of the valve when the engine is hot.

タペットとバルブあるいはプッシュ・ロッドとの間には，エンジンが熱いときにバルブが若干膨脹するのを**考慮**して，バルブが適切に閉じることが**できる**ように常に隙間を残している.

can : auxilary verb expressing ability or knowledge of how to do something or permission or desire or liberty to act	**できる**

Only when you use the proper cut-off wheel, you **can** get the best results.

最良の結果が得**られる**のは，適切な切断砥石を使ったときだけである.

Greater oil flow would be obtained if the oil-supply hole **could** be placed at the point of maximum clearance.

給油孔を隙間が最大の位置にすることが**できれば**，油流量はもっと大きくできる.

A tool **can** be operated by hand.

道具は，手で操作**できる**.

Scissors with a carbon content of up to 0.35 percent **can** be produced economically by cold forming.

炭素含有量0.35%までの鋏は，冷間成形で経済的につくることが**できる**.

The material **can** rarely be used for cutting tools.

この材料を切削工具に使うことが**できる**のは稀である．

Continuous CO lasers **can** produce deeper welds at higher rates of speed than those **possible** with the Nd : YAG laser.

連続波 CO レーザは，Nd：YAG レーザで**可能な**速度よりも高速で，より深い溶接をすることが**できる**．

A choice of spindle types is **available** and a 12-station automatic toolchanger **can be** added as an option.

主軸形式は選択**可能**で，またオプションで12ステーションの自動工具交換装置を追加**できる**．

---

capable : competent, having a certain ability or capacity

できる，能力がある

A battery of 40 amp-hour **capacity is capable of** discharging at the rate of 1 amp. for 40 hours, or 4 amp. for 10 hours, and so on in proportion.

容量40アンペア時のバッテリは，40時間1アンペア，10時間4アンペアなどの割合で放電することが**できる**．

Tool changeovers **can** be accomplished in 6 seconds. The machine **is capable of** producing a complete part every 100 seconds.

ツール交換が6秒で**できる**．この機械は，100秒毎に完成部品をつくることが**できる**．

The Model D824-12 surface grinder is said to **be capable of** a variety of surface grinding. With the proper accessories, cylindrical and form grinding operations on a production basis **are possible**.

このモデル D824-12平面研削盤は，いろいろな平面研削が**できる**という．適切な付属品を使えば，生産ベースで，円筒研削および総形研削が**可能である**．

The N is an extremely versatile machine with **capability** of efficient and accurate sharpening of cutters and tools.

Nは，カッタおよびバイトを効率良く正確に研ぐことが**できる**きわめて汎用性のある機械である．

The grinder **is capable of** grinding cutting tools without special accessories. Grinding **capabilities** include ball, taper, …… ; reamers.

この研削盤は，特殊な付属品を使うことなく切削工具を研削**できる**．研削**できる**のは球，テーパ，……，リーマなどである．

capacity : the ability to contain or accommodate, ability, capability　　**できる（容量，能力）**

Model T-125 **has the capacity** for cold forming solid pins, shafts, tenons, and rivets up to 1/8″ diameters, larger sizes from tubular stock.

T-125形は，径1/8インチまでのムクのピン，軸，ほぞおよびリベットを冷間成形**できる**．1/8インチより大きい寸法はパイプ材からつくる．

It **is able to** perform a series of functions and **has the capacity** to change quickly and easily to an alternate pattern with little or no retooling.

一連の機能を果たすことが**でき**，ほとんどあるいはまったくツーリングし直さないで，これに代わるパターンに迅速かつ容易に変換**できる**．

Programs **can** be stored on a small magnetic tape cassette, with **capacity** to accept up to 32 part programs, that **can** be edited as required.

プログラムは，最大32種のパーツ・プログラムを受け入れる**能力**をもった小さい磁気テープ・カセットに記憶することが**でき**，そして必要に応じて編集**できる**．

Tapping **capacity** from 10/32″ up to 1½/16″ in mild steel.

ネジ切り**能力**は軟鋼で10/32インチから1½/16インチまで．

The machine has a 14″ swing over the table, and a 30″ **capacity** between centers. It has a hydrostatic headstock for precision grinding, with the **capability** of holding roundness within 0.000010″ and

この機械は，テーブル上の振り14インチで，両センタの**容量**は30インチ．真円度を0.000010インチ以内に保ち，滑らかさが3 μインチの表面仕上げをつ

producing surface finishes as smooth as 3 μin.

くる**能力**がある精密研削用静圧主軸台を備えている．

This drillhead has a 5/32″ chuck. Its **capacities** are 1/8″ diameter in brass.

このドリル・ヘッドには，5/32インチのチャックが付いている．その**能力**は，真鍮で直径1/8インチである．

CNC, rotary tables with 6 to 100 ton load **capacities**.

積載**能力** 6 ～100トンの CNC ロータリ・テーブル

enable : to give the means to do something　**できる，手段を与える**

The coating **enables** the tiles to withstand temperatures up to 2,300°F.

このコーティングによって，タイルは2,300°Fの温度まで耐えることが**できる**．

It is the operation of this principle which **enables** one of the water-cooling systems of car engines to function.

車のエンジンの水冷方式のうちの１つが機能**できる**のは，この原理の働きである．

Off-set jaws **enable** the spanner to be used in a confined space.

オフセット・ジョーによって，スパナは狭いところでも使うことができる．

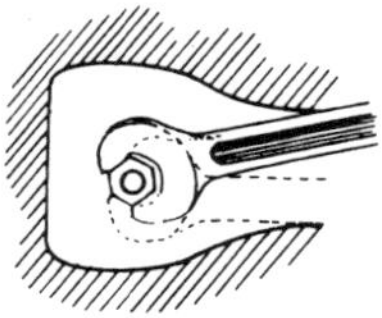

The ease and simplicity of operation of the D **enable** relatively unskilled operators to produce parts to close limits without expensive tooling.

Dの操作は，簡単かつ容易であるから，比較的未熟な作業者でも，高価なツーリングを使うことなく狭い公差で部品をつくることが**できる**．

The gears are illuminated from below, **enabling** the operator to see that they are machining correctly and free from shock.

歯車装置は下から照明されているので，作業者は歯車が正しく切削されていて衝撃のないことを見ることが**できる．**

Visual alarms **enable** the operating engineer to perform other tasks in the vicinity of the control center without requiring his constant supervision.

警報装置が見えるので，操作員はいつも監視している必要がなく，コントロール・センターの近くで他の仕事をすることが**できる．**

This package **enables** diagnostic tapes to be read via the standard control tape reader.

このパッケージは，診断テープを標準の制御テープ・リーダで読むことが**できる．**

The gearing **enables** screws of varying pitch and diameter to by varying the speed of rotation of the lead screw.

この歯車装置で，親ネジの回転速度を変えることによって，いろいろ違うピッチおよび直径のネジを切ることが**できる．**

Fortunately all modern carburettors are provided with adjustments, so that the maker **is enabled** to adjust the instrument to the particular car.

幸い，今日の気化器はすべて調整装置を備えている．したがって，メーカーは装置を特定の車に調整**できる．**

fail: to be unable to do something　　**できない，うまくいかない**

The tool **fails** to cut satisfactorily.

バイトが，よく切れ**ない．**

Work hardening **fails** to increase the resistance of material to wear.

加工硬化で，材料の耐摩耗性を上げることは**できない．**

If a catalog **fails** to provide this information, it **can** be approximated by averaging the bearing ID and OD.

もし，カタログでこの情報が得**られない**場合には，軸受の内径と外径を平均することによって概算**できる．**

The more obvious reason for a variation in output is a loose driving belt ; this might drive the dynamo at low speed, but **fail** to do so when engine speed rises.	出力変動のよりはっきりした理由は，駆動ベルトが緩いことである；これではダイナモが低速のときは駆動できるかもしれないが，エンジンの速度が上がると駆動**できない**．
If, after this has been done, the petrol still **fails** to flow freely, the pipe line itself must be removed and examined for a blockage.	もし，これをやっても燃料がまだ自由に流れ**ない**なら，パイプ・ライン自体を取外して，詰まりを調べなければならない．
The cut-out points **fail** to make good contact due to their being pitted and dirty.	切り離し点は，小穴ができかつ汚れているため，うまく接触**できない**．
Failing this the radiator will require raising a little.	これが**駄目**なら，ラジエータをわずか上げることが必要になろう．

feasible : able to be done, possible　　**(することが) できる**

The measurement of cage motion using proximity probes was difficult but proved to **be feasible**.	近接プローブを使って，保持器の運動を測定することはむずかしかったが，**できる**ことはわかった．
Without altering the mechanical aspects of the system, a couple of ways exist to reduce the amount of mist. One is to increase the viscosity of the oil (if this **is feasible**) and the other is to use antimist additive.	システムの機械的な部分を変えないで，ミストの量を減らす２つの方法がある．１つは油の粘度を上げること（もし，これが**できる**なら），そしてもう１つはアンティミスト剤を使うことである．

give: to permit a view of or access to	できる

The extra-heavy columns and cross rail **give** the support for the horse power needed to drive modern cutters to full capacity.

特別強いコラムとクロスレールは，新式のカッタを能力一杯まで駆動するために必要な馬力を支えることが**できる**．

Continuous CO_2 lasers **give** much faster cutting speed with very good edge quality.

連続波 CO_2 レーザによって，切断面の品質が非常に良く，しかも，非常に早い切削が**できる**．

The bench vise is a basic but very necessary tool in the shop. With proper care and use, this workholding tool will **give** many years of faithful service.

ベンチ・バイスは，工場の基本道具で，必要性の高い道具である．適切な注意と使い方で，この工作物保持具は長い年月，安心して使うことが**できる**．

may (**might**): expressing possibility or permission	できる，してもよい

Next, lever the valve stem through its washer so that a grip **may** be obtained on its head to withdraw it.

つぎに，弁を引き出すのに弁かさがつか**める**ようにワッシャを介して弁棒をこじり上げる．

The maximum diameter which **may** be ground is dependent on the external dimensions of the component which **can** be effectively gripped in the workholding fixture.

研削**できる**最大直径は，ワーク保持具が実際につかむことの**できる**外部寸法による．

permit: to make possible	できるようにする

A keyboard entry control panel **permits** machine programming and cutting regulation.

キーボード入力方式制御パネルで，機械のプログラミングおよび切削を調整**できる**．

An articulated arm **permits** lifting, turning, and placing of heavy and awkward-shaped workpieces into restricted work zones.

関節腕によって，重くて面倒な形の工作物を限られた作業域内に持上げ，回わし，そして置くという作業が**できる**.

A floating main drive pulley **permits** headstock travel with the spindle.

浮動式主軸駆動プーリは，主軸台が主軸と一緒に動くことが**できる**.

Unit construction of the power elements **permits** their removal without disassembling the entire machine.

動力部の要素部品はユニット構造であるので，機械全部を分解することなく，取外し**できる**.

The heavy duty bearing has a relatively thick and rigid outer race **permits** its use in a split housing.

重荷重用軸受は，比較的厚く，外輪が高剛性であるから，それを割り形ハウジングに使うことが**できる**.

The toolchanger is a stand-alone unit. It **permits** a 25-second tool-to-tool time.

ツール・チェンジャは，独立形ユニットで，ツールからツールまで25秒**ですむ**.

All bearings are permanently lubricated. This bearing design and the elimination of a moving quill **permit** very high radial and thrust loads, and facilitate chatter-free milling over the speed ranges.

軸受はすべて無給油式である．この軸受設計とムービング・クイルの排除によって，きわめて高いラジアルおよびスラスト負荷が**可能**で，かつ全速度範囲にわたり楽にビビリのないフライス加工ができる.

DoALL power saws are designed to **permit** higher band tension for more accuracy, **allow** higher feed rates, and **provide** greater work efficiency and less operator fatigue with centralized controls.

D社の動力鋸は，精度を上げるため帯鋸の張力を高く，送り速度を早く，集中制御装置を使って作業効率を上げ，かつオペレータの疲労を軽く**できる**ように設計してある.

The gearbox **permits** higher and lower speeds than those of the prime mover, **allows** "hot shifting" at any point within the selected speed ranges, and **can** be easily set up for automatic shifting at points that give the greatest efficiency.

歯車箱は，原動機の速度より高速あるいは低速に**でき**，また選定速度範囲内のどんな点にも即座にシフト**でき**，かつ，最大効率が得られる点で自動シフトするよう容易に段取りすることが**できる**.

An electromagnetic brake is designed to stop the spindle fast, and hydraulic clutches help **permit** the spindle to reverse quickly and easily.

電磁ブレーキは，スピンドルを迅速に停止するよう設計されていて，油圧クラッチの助けで，スピンドルの逆転が迅速かつ容易に**できる**.

The UI vision system **permits** U robots to assemble, inspect, and perform metal handling functions.

このUI視覚システムによって，ロボットは組立，検査，および金属のハンドリング機能を果たすことが**できる**.

The patented Unisorb Fixator System **permits** alignment adjustments to be made after anchor nuts are tight. Thus, tolerance to tenths (0.0001″) **can** be achieved without trial-and-error jacking and tightening during installation.

この特許のUFシステムは，アンカー・ナットを締めてから，アライメントの調整が**できる**.それで，据付け中にジャッキングおよび締付けの試行錯誤することなく0.0001インチ桁の許容差で据付けることが**できる**.

New Positive Depth Control **permits** the self-feed, or pull-out, of a tapping spindle to be set from 1/16″ to 1/2″.

新しいPPCは，タップ立て，スピンドルの自動送り，あるいは引き出しを1/16インチ～1/2インチにセット**できる**.

It is the first programming system that **permits** operators with absolutely no knowledge of NC programming to generate a program-merely by pressing buttons.

これは，NCプログラミングの知識がまったくないオペレータが，――ボタンを押すだけで――プログラムをつくれるようにした最初のプログラミング・システムである.

Flow control valves **permit** precise adjustment of stroke speed; adjustable actuating arm controls stroke length. Foot Switch **permits** operator to use both hands on the work piece.

流量制御弁によって，ストローク速度を精密に調整**できる**；可変式作動アームはストローク長さを制御でき，フット・スイッチであるから，オペレータは両手を使って部品を扱うことが**できる**.

The purpose of removing the pistons is generally to fit new rings and/or pistons or to **permit** of re-boring.

一般に，ピストンを取外す目的は新しいリングおよび（または）ピストンを取付けるか，あるいは再中ぐり加工が**できる**ようにすることである.……

Exhaust valve.——The cylinder valve of the engine which is made to open once for every two revolutions of the crankshaft in order to **permit** of the escape of the burnt mixture.

排気弁.——燃焼した混合ガスが逃げ**られる**ように，クランク軸の2回転に1度開くようにしたエンジンのシリンダ弁.

possible: capable of existing or happening or being done or used etc.	できる

This check **is** seldom **possible** with ball bearings.

このチェックは，玉軸受についてはほとんど**できない**.

It is usually **possible** to tap the shaft gently in an endwise direction from each end alternately.

通常，軸へ各端から軸方向に交互に，ゆっくりタップを立て**ることができる**.

Full-film lubrication under very high load, very slow-speed conditions is **possible** by using hydrostatic lubrication.

静圧潤滑を使うことにより，高負荷できわめて低速状態の完全油膜潤滑が**可能である**.

It is possible to do this also with a cassette tape, but at any serious level of operation it is probably better to use disks, although they are much more expensive.

これはまた，カセット・テープを使って**することができる**が，重要な作業水準のときには，たいへん高価であるがディスクを使うほうが多分ベターである．

With small cubic components **it is possible to** machine pockets or internal surface using an end-milling cutter as shown in Fig. 12.

小さい立方体部品は，図12に示すように，エンドミルを使って，ポケットあるいは内面を切削**できる**．

It is possible that a cutting tool is held …….

バイトを……に保持する**ことは可能である**．

Since each step of the procedure is in numerical sequence **it is possible to** reduce the chance for human error.

作業の各ステップは数字順であるから，人が間違える可能性を少なくする**ことができる**．

Rotation in the opposite direction is also **possible** provided the motion is not oscillatory in nature.

運動の性質が揺動でないものとすれば，反対方向の回転もまた**可能である**．

The coating increases the tool's surface hardness and friction resistance, thus **allowing** higher cutting speeds. This **is made possible** because of the elimination of cobalt on the cutting surface and the addition of new materials such as nitrides and oxides.

コーティングはバイトの表面硬さと耐摩擦性を上げ，これでより切削速度を早く**できる**．これは，切削上面の Co をなくし，かつ窒化物および酸化物のような新しい物質を加えることで**可能になった**．

Many lathe operations would **not be possible** without the use of the steady and follower rests. These valuable attachments **make** internal and external machining operations on long workpieces **possible** on a lathe.

旋盤作業の多くは，固定振れ止め，および追随振れ止めを使わなくては**できない**．これら価値ある付属装置によって，長いワークの内側・外側の切削作業が旋盤で**できるようになる**．

Reseach at Cincinnati Milacron **has made it possible to** successfully broach both thin-wall ferrous castings and alminum castings.

C社での研究により，薄肉の鉄鋳物およびアルミ鋳物のブローチ加工がうまく**できるようになった．**

The accuracy of the headstock and the use of high-quality precision bearings **make** it **possible** for the company to grind heavy workpieces with no problems.

主軸台精度の改善と高品質精密軸受の使用によって，この会社は重い工作物をまったく問題なく研削**できるようになった．**

……have **as** equal a flow **as possible** of polymer.

……は，**できるだけ**ポリマと同じ流量にする．

To minimize the possibility of strain cracking, hardened parts should be tempered **as** soon **as possible** and not left at room temperature for long periods.

歪亀裂の可能性を最小限にするため，焼入部品は**できるだけ**すぐ焼戻し，長期間室温で放置しない．

Apply thinnest **possible** film of A4 Metal set to face of tip block to fill crack.

亀裂を埋めるため，チップ・ブロックのすくい面に(つける)Aメタルを**できるだけ**薄い膜にする．

The X-and Y-axis **possible** positioning accuracy is ±0.0002″; **possible** X-, Y-and Z-axis repeatability is ±0.0001″

XおよびY軸の**可能な**位置決め精度は±0.0002インチ；**可能な**X, Y, Z軸の繰返し精度は±0.0001インチである．

provide: to cause (a person) to have possession or use of something

できる，～が得られる

……, thus **providing** easy access for cleaning.

これで，洗浄のために容易に近付くことが**できる．**

The grease **is able to** withstand high temperature and **provides** proper lubrication.

グリースは，高温に耐えることが**でき**，かつ適切な潤滑が**保てる．**

ととのえる（調製，用意，準備）

improvise: to compose (a thing) im-prompt : to provide in time of need, using whatever materials are at hand	（即席で）つくる

An **improvised** method of removing a pulley or similar part is illustrated, but leverage should be applied with discretion.

プーリやこれと似た部品を取外す**即席でできる**方法を示すが，こじるのは慎重にすること．

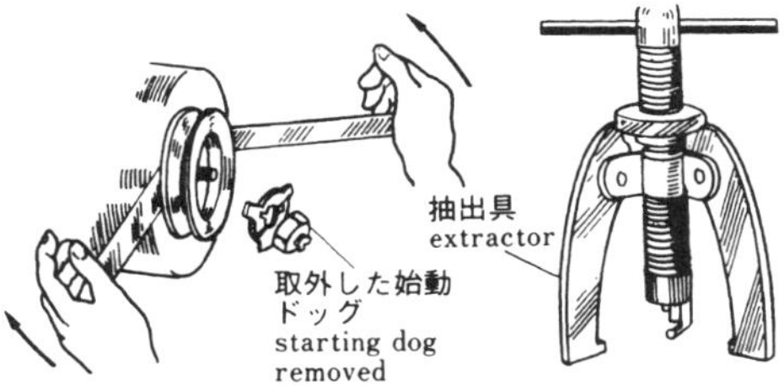

In working out kinks in the rather difficult area near the channel of the wing an **improvised** swage **made** from a large sawn-off tyre lever will come in useful.

ウィングのチャンネル近くにあるかなりやりにくい部分のねじれを直す場合，鋸で切った大きなタイヤ・レバーで**つくった即席の**スエージ〈金型〉が役立つようになる．

prepare: to make (food or other sub-stances) ready for use	つくっておく，調製する

Prepare a patch of T3 aluminum alloy, 0.020 inch thick.

T-3アルミ合金のパッチ（厚さ0.020インチ）を**つくる．**

A final plan **is prepared**.

最終案を**用意する．**

In **preparing** these samples, one shield was removed and 10 microliters of oil were applied from a syringe along the race.	これら試料の**調製**には，シールを１つ取外して，軌道輪に沿ってスポイトで10$\mu\ell$の油を入れた．
The test bar SAE 1020 (0.2 percent carbon) hot rolled steel with a Brinell hardness number (Bhn) of about 120, **are prepared** according to the specifications in Fig. 2 .	試験用のバー（ブリネル硬さ(Bhn)約120のSAE1020(0.2%炭素)熱間圧延鋼)を，図２の仕様に従って**つくった．**
C coating, **prepared** to meet MIL-L-46010 specification, …….	MIL-L-46010の仕様に合うように**つくった**被膜Cは…….
A tape head cleaning system maintains clean heads and rollers in NC tape **preparation** systems.	テープ・ヘッドのクリーニングシステムは，NCテープ**作成**システムのヘッドおよびローラをきれいに保つ．
Carbide inserts are available with several different edge **preparations**.	超硬インサートは，**既製の**いくつか違ったエッジのものがある．

取る（面を）

chamfering ; chamfer : to bevel the edge or corner of.	**面取り，食付き部**（タップ，チェーサ）

Chamfer workpiece before using die.	ダイスを使う前に，工作物を**面取りする．**
Chamfer the end of the thread to protect it from damage.	ネジを損傷しないように，その端を**面取りする．**

With special tool parts may **be chamfered,** formed, grooved or flanged.

特殊なバイトで，部品の**面取り**，形状，溝，フランジを切削できる．

Chamfering is accomplished in a similar manner.

同様の方法で，**面取りする．**

〈用　語　例〉

bearing chamfer　　軸受面取り

gear tooth chamfering machine　　歯車面取り盤

---

round : to make or become round　　**丸面取り，丸める**

It should be noted that the **rounding** of the corner of the basic rectangular shape is ignored in deciding the geomertic code.

ただし，この寸法・形状のコード決定では，基本長方形の隅の**丸面取り**を無視している．

Make sure the wrench you select fits properly. If it is a loose fit, it may **round**

選んだスパナがぴったり合うことを確める．もし，はめあい

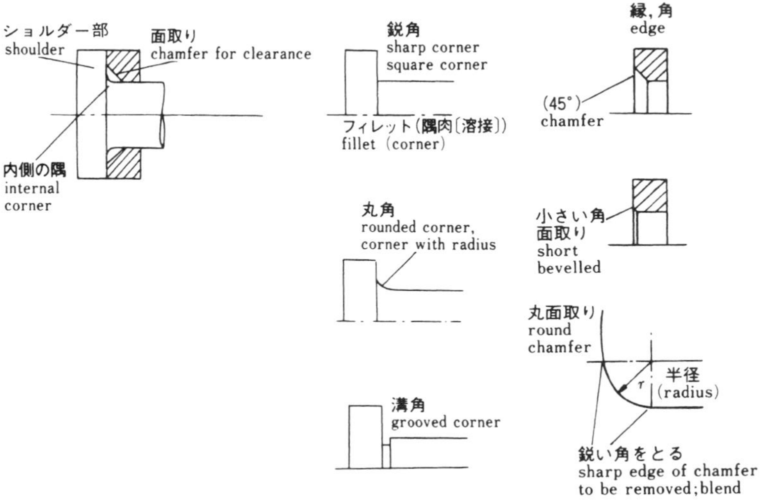

隅，縁，角
corner, edge

off the corners of the nut or bolt head.

が弛いと，ナットやボルトの頭の角を**丸くしてしまう**ことがある．

取付ける，置く，降ろす，乗せる

attach : to fix to something else　　**取付ける**

Fix one end and **attach** a weight to the other end of the thread.

糸の一端を固定して，他端に錘りを**取付ける**、

The dynamometer **was attached** to the fixture and the fixture clamped to the bed of a universal testing machine.

ダイナモメータを取付け具に**取付け**，この取付け具を万能試験機のベッドに固定した．

To this central system various NC and CNC controls may **be attached**.

この中央システムには，いろいろなNCおよびCNC制御装置が**取付けられる**．

Prop the window open while the screws by which it **is attached** to the frame of the window are removed.

窓枠に窓を**取付けている**ネジを取外す間，窓を棒で支えて開いておく．

Cap-nut : A nut which covers the end of the bolt to which it **is attached**, thus giving a better appearance to the finished part than an ordinary nut.

袋ナット：ボルトに**取付けて**，その端を覆うナット．これで普通のナットよりも仕上部品の見ばえがよくなる．

Attach…horizontally at 20cm height from the ground surface.

…を地面より高さ20cmに水平に**取付ける**．

The **attaching** and **detaching** of parts ….

部品の**取付け・取外し**は……．

A maintenance schedule should **be**

保全スケジュールには，再給

attached to the report giving details of relubrication, inspection routines, operating temperature, etc.

油，検査手順，運転温度などの細目を記入した報告書を**添付**のこと.

deposit : to lay
: to put down

置く，降ろす

Both tools **are deposited** in the storage position, and two new tools are then inserted in the spindles.

両方のツールを格納ステーションに**降ろし**，2つの新しいツールをスピンドルに挿入する.

A rotating arm then swings into place at the pickup position, descends and picks up the part, elevates and swings into the die area where the workpiece **is deposited**.

つぎに，回転アームがピックアップ位置の所定位置で旋回し，下降して部品をつまみ上げ，上昇して型領域内で旋回し，そこで工作物を**降ろす**.

The total cycle time is six seconds, made up of a 0.75 sec. index, 0.62 sec. for raising yoke to clamp, 4 sec. for units to oparate and 0.62 sec. for yoke to lower and **deposit** parts **on** the transfer bar.

全サイクル時間は6秒．その内訳はインデックス0.75秒，ヨークを上げてクランプするのに0.62秒，装置の作業に4秒，ヨークを下げてトランスファ・バーにこの部品を**置く**のに0.62秒である.

A M robot with dual grippers exchanged a blank workpiece with a finished one at the machine chuck, then transferred the workpiece to a gaging station, and then **deposited** the workpiece either **on** an "accept" conveyor or a "reject" conveyor according to feedback from the gaging station.

複式グリッパのMロボットは，機械チャック位置で粗材と仕上工作物を交換し，つぎに工作物を測定ステーションに移送する．つぎに，測定ステーションからのフィードバックに従って，合格コンベヤまたは不合格コンベヤのどちらかの**上に**，工作物を**降ろす**.

Parts catcher **deposits** finished part

部品キャッチャは，仕上部品

behind access door in chip guard.

を切屑よけの出入扉の後に**置く**．

The processing cycle requires the robot to find the material on the input conveyor, place it on a saw, then transfer the cut piece through the wash, dry, and etch processes and **deposit** the part **at** an output station.

この処理サイクルは，ロボットが入力コンベヤ上のものを見付け，それを鋸盤に置き，ついで，切断したものを洗浄，乾燥，エッチング工程に移送し，出力ステーション**で**，その部品を**降ろせる**ことが必要である．

……. The second gripper takes the part and rotates. The now-inverted part **is** then **deposited** in the second chuck, ready for machining.

……．第2のグリッパが，部品を取って回転する．つぎに，この逆さにしたばかりの部品は，すぐ切削できるように，第2のチャックに**取付けられる**．

fit : to put into place　　**取付ける，はめる**

Faulty seatings, burrs, dirt, and similar causes, prevent correct ring expansion. It is then necessary to dismantle the parts, correct the fault, and **re-fit**.

欠陥のある座，バリ，ごみなどがあると，それが原因でリングは正しく膨張できない．そういうときには部品を取外し，欠陥を修正して，**取付け**直すことが必要である．

When a replacement bearing **is fitted**, new shims will probably be required.

代わりの軸受を**取付ける**ときには，多分新しいシムが必要になる．

Bearings **are**, usually, **fitted** in their housing after being placed on the shaft.

軸受は通常，軸に取付けてから，ハウジングに**取付ける**．

The milling cutter is mounted on a shaft called an arbor, whose extremity **fits** into a tapered socket in the driving spindle.

フライスカッタはアーバーと呼ばれる軸に取付けられ，アーバーの先端は駆動スピンドルのテーパ・ソケットに**はめ込む**.

……with a screw driver which **fits** perfectly into the slot.

（ネジの）すり割りに完全に**はまる**ネジ回しで…….

If **fitted** to the nut in the opposite way the spanner will be damaged and the grip will be weakened.

もしナットに逆に**はめる**と，スパナが損傷し，把握力も弱くなる.

The test assembly **fits** onto the upper end of vertical drive shaft which is driven at 1,500rpm by a 3/4 Hp motor.

試験組立品を，3/4馬力のモータが1,500 rpmで駆動する立駆動軸の上端に**取付ける**.

Sleeves **fit** snugly over the workpiece.

スリーブは，工作物にしっかり**はまる**.

A spacer **is fitted** between bearing inner ring and the shoulder.

スペーサを軸受内輪とショルダー部との間に**取付ける**.

ZZ: Shield **fitted** at both sides of bearing.

ZZ：軸受の両側に**取付けた**シールド.

The complete bearing **is fitted** square with the shaft axis.

軸受全体は軸心と直角に**取付けられている**.

An excellent and easy method of **fitting** an inner ring is to put the shaft (if short enough) under the ram of a hand-screw mandrel press, using a tubular sleeve to transmit the pressure to the inner ring.

内輪の，簡単でうまい**取付け**方法は，内輪に圧力を伝えるためにチューブ状のスリーブを使い，軸を手動スクリュ・マンドレル・プレスのラムの下に（短かくて置けるなら）置くことである.

Self-aligning bearings should swivel easily in all directions. With roller bearings it is usually possible to check the diametric

自動調心軸受は，あらゆる方向に容易に旋回する．コロ軸受は，**取付け**後，転動体と外輪軌

slackness after **fitting** by inserting feelers between the rolling elements and the outer -ring track, lifting the outer ring mean while to take the weight off the feeler gauge.	道の間に隙間ゲージを挿し込んで（この間，隙間ゲージに重みがかからないように外輪を持上げて），直径すきまをチェックすることが，通常可能である．
If the bearing has an **interference fit** on the shaft, a puller should be used.	軸受が軸に**しまりばめ**であるなら，プラー（引抜き工具）を使うこと．
The inner ring of bearings with a tapered bore is always mounted with an **interference fit** usually on an adapter sleeve or a withdrawal sleeve.	テーパ穴付軸受内輪は，通常アダプタ・スリーブあるいは取外しスリーブに，常に**しまりばめ**で取付ける．
The spline **fit** (circular clearance) is 0.20～0.25 mm.	スプラインの**はめあい**（円周方向隙間）は，0.20～0.25mm．
The **fit** of washer (rotating ring) of a thrust bearing on the shaft is not quite so important as the corresponding **fit** for a radial bearing.	スラスト軸受のワッシャ（回転輪）の軸との**はめあい**は，ラジアル軸に対する**はめあい**ほどには重要ではない．
After the bearing has been in operation under load, the nut should be tighten up slightly to take up any looseness in the **fit** that may have developed.	軸受を負荷運転したら，**はめあい**の緩み（生じていることがある）をすべて取り除くように，ナットをわずかに締め上げる．

〈用　語　例〉

clearance fit　隙間ばめ
interference fit　しまりばめ
transition fit　中間ばめ

thread plug gauge for checking fit
はめあい点検ネジプラグ・ゲージ

fit up: to supply or equip　　**設ける，取付ける**

The spindle adapters for the A-2 **fit up** into the lower bearings of the spindles and	A-2用の主軸アダプタは主軸の下部軸受の中に**取付けて**，

offer positive retention by the locking nut. A spindle brake is provided for the spindle in the operating position to facilitate tool loading.

固定ナットで確実に止められている．主軸には，作業位置で楽にツールを装塡できるように，主軸ブレーキが付いている．

install: to settle in a place, to set (apparatus) in position and ready for use　　**取付ける，据付ける**

Oil leaks can occur at the filter housing if the filter adapter gasket **is** not **installed** properly, or if excessive torque is applied to the attaching bolt.

フィルタ・アダプタガスケットが適切に**取付けられ**ていなかったり，(締付け)トルクが取付けボルトにかかりすぎていると，フィルタ・ハウジングに油漏れを生じることがある．

Prior to **installing** a bushing with the arbor press, what two important steps must be taken?

アーバー・プレスでブッシュを**取付ける**のに先立ち，踏むべき2つの重要な手順は？

Temporarily **install** retaining nut on fitting.

取付け金具に，固定用ナットを仮に**取付ける**．

The Oil-Mist **installed** on top of 2 tandem mill …….

2つのタンデム圧延機の頂部に**取付けた**オイル・ミスト（装置）は…….

Install new O-ring in groove in top of transmission case.

ミッションケース上面の溝に，新しいOリングを**取付ける**．

Install new stud to project 10～12 mm above the surface.

新しい植込みボルトを，表面から10～12mm突出るように**取付ける**．

Install …… in the reverse order of removal.

取外しの逆の順序で，……を**取付ける**．

Careful **installation** and removal of stylus assembly should **be accomplished** while protected by the stylus guard.	演奏用針組立品は，針ガードで防護したまま，注意して**取付け**・取外し**すること**．
In some **installation**, the rolling elements are carried in a flexible cage similar to a roller chain.	**取付け方法**には，転動体をローラチェーンに似たフレキシブルな保持器に納めたものもある．
lay : to place or put on a surface or in a certain position	**置く**
The bearing should be removed from its protective wrapping just prior to fitting. It should **be laid on** this wrapping and not directly on the bench or the shop floor.	軸受は，取付ける直前に，保護用包装から取出す．軸受は，この包装の上に**置き**，じかに作業台や工場の床の上に置かないこと．
If assembly is not in balance, **lay** small washers **in** base to attain balance.	組立品がバランスしていない場合には，バランスするようにベースの中に小さいワッシャを**置く**．
A Straight-edge **is laid** across the boss of the adjustment gauge and the bearing surface of one lever ; the adjustment nut is then turned until the lever touches the underside of the straight-edge.	直定規（ストレッチ）を，調整ゲージのボスと１つのレバーの支持表面に渡して**置く**．つぎにレバーがストレッチの下側に触れるまで，調整ネジを回わす．
locate : to assign to or establish in a particular location, to be situated there	**位置決めする，～にある，取付ける**
The block **is** automatically **located** in a fixture.	ブロックは，取付け具に自動で**取付けられる**．

The automatic tool changer **is located** on the column and away from the work area.	自動工具交換装置は，作業領域より離してコラムに**取付けられている．**
The spindle selector, which indexes with the turret, **is located** on the top side within easy reach of the operator.	タレットでインデックスするスピンドル・セレクタは，作業者が楽に手の届く上面に**取付けられている．**
Operating buttons, automatic oiler, ……, and indicators **are located** at the right front corner of the machines where easily and quickly reached.	操作ボタン，自動給油器，……インジケータは，楽にかつ素早く手の届く機械の右前隅に**ある．**
Two spherical roller bearings, which **are** not **located** axially, are used to support the radial loads.	軸方向に**位置決め**されてない2つの自動調心コロ軸受が，ラジアル荷重を支持するのに使われている．
The spindle unit then **relocates** to the tool change position.	つぎに，スピンドル・ユニットはふたたびツール交換**位置になる．**

mount : to put into place on a support, to fit in position for use or display or study — **取付ける**

Mount the separable ring first and oil its raceway slightly. After oiling or greasing the rollers, fit the other ring with roller and cage assembly.	最初に，分離できる軌道輪を**取付けて**，その軌道に油を少し塗る．コロに油またはグリースを塗ってから，保持器付コロの付いたもう一方の軌道輪を取付ける．
…… **is** normally **mounted** with zero clearance.	……は普通，隙間をゼロに**取付ける．**

Most bearing applications have a rotating inner race and stationary outer race, so the inner race should **be mounted** to the shaft or adapter ring with a press fit and the outer race should **be mounted** to the housing with a slip fit.

軸受は，内輪回転，外輪静止で使われることが多い．そこで，内輪は軸あるいはアダプタ・リングにしばりばめで**取付け**，外輪はハウジングにすべりばめで**取付ける**．

The model I turret head **is mounted** to the turret slide by means of a large diameter ball bearing.

I形タレット・ヘッドは，大径玉軸受によって，タレット・スライドに**取付けられている**．

…… **is mounted** against to the nose of spindle.

……は，主軸端に当てて**取付けられている**．

The tool holder **was mounted** in a vertical slide with manual adjustment.

ツールホルダを，バーチカル・スライド（上下すべり台）に，手で調整して**取付けた**．

Mount the workpiece in a bench vise so that the hole is in a vertical position.

穴が縦になるように，工作物を卓上バイスに**取付ける**．

The main spindle and the gears are all **mounted** in the headstock.

主軸および歯車は，すべて主軸台に**取付けられている**．

The duplex-pair test bearings **were mounted** on the shafts and the outer race pairs were fitted into free-floating housings.

試験用の組合軸受を軸に**取付け**，その一対の外輪を自由浮動するハウジングにはめ込んだ．

Work table **on** which the work **is mounted** …….

工作物を**取付ける**ワークテーブルは…….

Mounted on the guideways of the lathe bed **is** the saddle or carriage, which is constructed as a compound slide.

旋盤ベッドの案内面には，複式スライド構造のサドルまたはキャリッジが**取付けられている**．

The push button control station **is mounted** in a convenient position on the spindle head and contains all the controls required for ease of machine operation.

押ボタン式制御ステーションは，主軸頭に使いやすい姿勢で**取付けられ**，機械操作を容易にするために必要なすべての制御装置が納まっている．

An electric motor **mounted** at the rear of the machine …….

機械の後部に**取付けられている**電動機は…….

All controls **are** front-**mounted** for safety and convenience.

すべての制御装置は，安全で便利なように前面に**取付ける**．

All CNC controls for the Series M **are** pendant-**mounted** for easy, total control of machine operation.

シリーズ M のすべての CNC 制御装置は，機械操作が容易，かつすべて制御できるよう，吊下げて**取付け**てある．

Machine controls **are** stanchion-**mounted** for convenience. The operator can easily and automatically locate and grind precision holes, radii, blends, and surfaces.

機械の制御装置は，便利な支柱**取付け**．精密な穴，隅の丸面，なめらかな接合面および表面を容易かつ自動的に位置決めして研削できる．

The model **is** caster **mounted** and said to be readily portable since no water connections are required.

この形式は，キャスタが**取付けられ**ており，水道につなぐ必要がまったくないから，容易に持ち運べるという．

Wheelhead with two wheels **is** swivel **mounted** at 30° angle on the machine base. Swivel mounting permits adjustment to make sure the part is ground straight.

砥石 2 枚を取付けた砥石台を旋回させて，機械ベースに30°の角度で**取付ける**．旋回取付けのため，部品の真直な研削が確実に行なえるよう調整できる．

These rollers can **be** shaft **mounted** on new conveyor lines.

これらのローラは，新しいコンベヤラインの軸に**取付けるこ**

とができる.

All are a fully enclosed design with recessed or flush-**mounted** air, electrical, and circulating oil controls, ……

エア，電気，循環油の制御装置などを，引き込むか面位置で**取付けた**完全密封設計.

Mounting should **be carried out** in a dustfree, dry environment.

ほこりや，湿気のない環境で，**取付け作業をすること.**

The most expert maintenance people have harmed bearing in **mounting** through inattention and carelessness,

保全のエキスパートでもほとんどの人が，油断したりうっかりして，**取付け**のときに，軸受を傷付けたことがある.

Front-**mount** plug-in design is now standard throughout the H. series, which simplifies **mounting** and makes installation quicker.

フロント**取付け**のプラグイン方式は，今はHシリーズの標準設計で，これで**取付け**が簡単になり，据付けも早くできる.

Normally delivered with four 7/16 holes for conventional **mounting**, it can be delivered undrilled for custom **mounting**.

普通は通常の**取付け**用の7/16インチ穴が4つあるものを荷渡しするが，顧客独自が**取付け**する場合には，穴明けしないで荷渡しもできる.

……**mountable** in any position without affecting efficiency.

……は，効率に影響を与えずに，どんな姿勢にでも**取付け可能.**

〈用　語　例〉

back-to-back mounting, indirect mounting　背面取付け
face-to-face mounting, direct mounting　正面取付け
mounting distance　組立距離
mounting machine　取付け機
mounted unbalance　装着のアンバランス
mounted wheel　軸付き砥石
outside diameter of mounted snap ring　止め輪の取付け外径
tandem mounting　並列取付け

place: to put into a particular place or rank or position or order, to locate	**置く，取付ける**

……. They **are** next **placed** on a belt conveyor. | つぎに，ベルトコンベヤの上に**置く**.

A shaft **is** then manually **placed** into a part nest on the pallet.	それから手で，シャフトをパレット上の部品巣箱の中に**納める**.
Vises should **be placed** on the workbench at the correct height for the individual.	バイスは，作業台にその人に合った高さに**取付ける**.
A drill sleeve **is placed** on the drill so that it will fit the taper in the tailstock spindle.	心押し台軸のテーパにぴったり合うように，ドリル・スリーブをドリルに**取付ける**.
Place the wheel in position and finally mount the outside bearing inner ring with its roller and cage assembly.	ホイールを所定位置に**取付け**て，最後に保持器付コロの付いた外側軸受内輪を取付ける.
Place a rubber bung in the container.	容器にゴム栓を**する**.
By **placing** a plastic bullet-shaped cap **over** the housing end during wheel assembly, the threads are protected, and assembly is facilitated. After assembly, the plastic tool is removed.	ホイール組立の間，プラスチックの弾頭形キャップをハウジングの端に**かぶせて**おくと，ネジは保護され，組立も楽になる. 組立が終わったら，このプラスチックの用具を取外す.
Place a drop of CCl_4 on the top of the workpiece.	工作物の上に CCl_4 を 1 滴**たらす**.
A chip is made of semiconductor	チップは半導体材料，それは，

material, usually, silicon, and is not in fact a chip of silicon but a rectangle formed from wafer thin slices of silicon **placed on** top of each other as a sandwich.	通常はシリコンでできているが，実際には一片のシリコンではなく，サンドイッチのように，それぞれが他の上面に**乗った**薄いシリコンの截片で形成された矩形のものである．
…… use microcomputers to program a robot arm for picking up and **placing down** objects.	……は，ロボット・アームが，物体をつまみ上げて下に**置く**ようプログラムするために，マイクロコンピュータを使っている．
When refitting care must be taken to see that they **are placed** with the polished side downward.	ふたたび取付けるときには，磨いた側を下にして**置かれている**のを，必ず目で確かめるよう気をつける．
If the maker's instruction book **has been misplaced** it is a good plan to make a copy of the fuse positions on a sheet of card which can be kept inside the lid of the fuse box.	もし，メーカーの取扱い説明書を**置き忘れて**いたら，ヒューズ箱のふたの内側に保管できるように1枚のカードにヒューズ位置を示すコピーを作るのも一案である．
It will be safe to **replace** the fuse **with** the spare provided in the clips on the frame of the mechanism.	ヒューズを装置の枠のクリップの中に用意してある予備品**と変える**ほうが安全．
All tool **placements** are automatically calculated at ±0.0008″ positional accuracy and stored in memory.	工具の**配置**はすべて，位置の精度±0.0008インチで，自動計算されて，記憶装置に記憶される．
The bench should **be placed** in such a position that it receives as much light as possible from the window.	作業台は，窓からできるだけ多くの光を受けられるような所に**置く**こと．

When making a seal of this kind, the tapered side of the groove should **be placed** to come nearest to the bearing it is to protect.	この種のシールを製作するときには，溝のテーパ側を，シールが保護すべき軸受に最も近くなるように**置く**.
The company introduced the M30 robot control based on programmable controller technology for load/unload, **pick/place applications.**	この会社の発表したM30ロボット・コントローラは，ロード・アンロードおよび**ピック＆プレース用**で，プログラム制御技術をベースにしている.
We develop a robot that can "handle the **pick-and-place** tasks and the press-feeding" tasks.	われわれは，**ピック＆プレース**作業，およびプレスの送り作業を処理できるロボットを開発した.
…… for the purpose of removing dirt or **placing** grease between.	ごみの除去，あるいは間にグリースを**詰める**目的で…….
Use a screwdriver or blunt chisel to lever the tacks out, **placing** the blade under the cloth and not directly under the heads of the tacks.	鋲をこじってとるのに，ネジ回し，または切れないたがねを使うが，刃を布の下に**入れて**直接鋲の頭の下には入れない.
Place one end of a wooden listening rod, screwdriver or similar object **against** the bearing housing as close to the bearing as possible, **place** the ear **against** the other end and listen.	木の聴音棒，ネジ回し，およびこれと似たものの一端を，できるだけ軸受近くの軸受ハウジングに**当て**，他端に耳を**当て**て，聴く.
Controls **are placed** above the machine, requiring less floor space.	制御装置は機械の上に**ある**から，必要な床面積は少なくてすむ.

position : the place occupied by a person or thing　**置く，位置決めする**
: to place in a certain person or thing

Essentially the same operation is performed in the manual mode, although in this case the operator must **position** the part at the pickup nest.

この場合，オペレータは部品を取出し巣箱に**置かなければ**ならないが，これと同じ作業は実質的には入手で行なわれる．

Reposition the workpiece so it is upright in the machine vise with the solid flange of the body up.

機械バイスに本体と一体のフランジを上にして，直立するように工作物の**姿勢をとり直す**．

The tools do not have to **be** located and **positioned** relative to each other.

ツールは，相互に相対的な位置および**姿勢を決める**必要はない．

The device has a rate of 3,600 cycles per hour and can **position** workpieces to within 0.001″.

この装置は，毎時3,600サイクルの速さで，工作物を0.001インチ以内に**位置決め**できる．

put : to move (a thing) to a specified place, to cause to occupy a certain place or position　**置く，取付ける**

Put the sample on the plate.

試料を皿の上**に置く**．

Put the cutter and spacing collar in place.

カッタと間隔筒を，所定の位置に**取付ける**．

Lock the nut and **put on** the hub cap immediately.

ナットを固定して，直ちにハブキャップを**取付ける**．

It is very important that these tools **not be put into** a drill chuck, because a burred

こういうツールは，ドリル・チャックに**取付けない**ことが非

shank can ruin a reamed hole as the shank is passed through it.

常に大事．というのは，バリのあるシャンクをリーマ仕上げの穴に通すと，穴を駄目にすることがあるからである．

The springs should certainly **be put** in order when repairing the covering, otherwise not only will the seat continue to sag but the covering will be subject to extra stresses.

カバーを修理するときには，バネを間違いなく，きちんと**置く**こと．そうでないと，シートがたるみっぱなしになるばかりでなく，カバーが余計な応力を受けることにもなる．

rest : to place or be placed for purpose.　**置く，静置する**

A ball **rests on** a plane surface.

球は平面上に**静置**されている．

The usual method of weighting the con-rod is to support one end at the same height as the scale pan, **resting** the other end **on** the scale ; the rod then is in a horizontal position and the little end can be weighted while the big end is on the support.

コンロッドの重さを測る普通の方法は，一端を秤量皿と同じ高さに支え，他端を秤りに**置く**．このときロッドは水平位置で大端部を支えに乗せておいて，小端部の重さを測ることができる．

The method of inserting the paper is a simple operation and can be performed rapidly. Hold the paper to be inserted with the left hand and place it behind the cylinder so that it **rests** lightly **on** the feed rolls——the small rollers between the paper table and the cylinder.

紙を挿入する方法は，簡単な作業で素早くできる．挿入する紙を左手で持ち，送りロール——紙のテーブルとシリンダの間の小さいローラ——に軽く**乗る**ようにして紙をシリンダの後に置く．

A spherical base **rests in** a corresponding seating in the housing.

球面の底部は，ハウジングのこれに対応する座の中に**据わる**．

The handle **is rested against** the compound. The chuck may be turned by hand.	ハンドルは，複合刃物台に**当ててとめる**．チャックは手で回わすことができる．

reset : to place again in some particular position ; to return a device to zero or an initial or arbitrarily selected conditions	**取付け直す，リセットする**

This table can **be** easily **reset** by the hand lever.	このテーブルは，ハンド・レバーで容易に**取付け直す**ことができる．
Feed the compound (rest) in 0.005 in and **reset** the cross feed dial to zero.	複式刃物台を0.005インチ送り込んで，横送りダイヤルをゼロに**セットし直す**．

〈用　語　例〉

reset group total　グループ別合計点検（グループ別合計精算）
reset to zero　帰零
reset total　合計精算
reset total key　精算キー
reset wheel for transaction　回数器帰零ホイール

ride : to sit on and be carried by (a horse etc.) : to be supported on	**乗る**

Strip of fluoroplastic anti-friction material are bonded to the table and saddle female way, which **ride on** hardened and ground inserted rectangular steel ways.	フロロプラスチックの耐摩擦材の細片がテーブルとサドルのへこんだ案内面に接着している．そしてこれは，焼入れ研削のインサート付矩形の鋼案内面に**乗っている**．
Two brass bearings **ride on** the journal of axle.	2つの真鍮の軸受は，アクスルの軸頸に**乗っている**．

set: to put or place, to cause to stand in position　　**取付ける，据付ける，設定する**

Set the gauge <u>by</u> <u>using</u> the bore indicator.	内径用インジケータを<u>使って</u>，ゲージを**セットする．**
……. Thus the gauge can only use to **set** the bearing with the aid of a gauge.	……．このように，このゲージは，あるゲージの助けをかりて軸受を**セットする**のにしか使えない．
For speeding up "<u>length</u> <u>setting</u>" of the turret tools.	タレット・ツールの，<u>長さ設定</u>のスピードアップに．
When a job is reordered, this gauge **is set** <u>to</u> previously recorded figures, and the respective tools **are set** <u>to</u> the gauge. The turret slide screw adjustment is then used to position one tool on the workpiece.	仕事を再受注したときには，このゲージを前に記録しておいた数字<u>に</u>**セットし**，そのゲージ<u>に</u>それぞれのツールを**セットする**．つぎに，1つのツールを工作物に位置決めするために，タレット・スライドのネジ調整を使う．
Set a indicator <u>at</u> <u>zero</u>.	インジケータを<u>ゼロに</u>**セットする．**
The compound **is** most often **set** <u>at</u> 14½ <u>degrees</u> to the right for right-hand external threads.	右（ねじれ）雄ネジは，複式刃物台を右<u>14½°に</u>**セットする**ことが多い．
Parting tool is clamped in the holder and should **be set** <u>on</u> or slightly above the center (<u>line</u>) of the workpiece.	突切りバイトは，ホルダに固定して，工作物の中心<u>線</u>あるいは少し上<u>に</u>**セットする．**
The compound micrometer collar should also **be set on** the zero mark.	また，複式刃物台のマイクロメータ・カラーを，そのゼロマ

ークに**セットする**.

Set fixture in their right place.

取付け具をその正しい位置に**取付ける**.

Set apparatus in position and ready for use.

装置を所定の位置にすぐ使えるように**据付ける**.

……by **setting** the gears that drive the lead screw to give the required pitch of the machined threads.

切削ネジが必要とするピッチが得られるように，親ネジを駆動する歯車を**セットする**ことによって……

Set the gauge so that the bore indicator again shows zero.

内径用インジケータがふたたびゼロを示すようにゲージを**セットする**.

All **setting are made** on direct reading dials, one **set** for horizontal and the other for vertical dimensions.

直読式ダイヤル（ゲージ）をすべて**セットする**．１つは水平方向の寸法用に，もう１つは垂直方向の寸法用に**セットする**.

All operations are completed at one **setting** of the workpiece.

工作物を一度**取付け**れば，すべての加工ができる.

Infeed **is set** by a microdial, with an accuracy of 0.0003″.

インフィードはマイクロダイヤルで精度0.0003インチに**セットする**.

The cutting speed and feed should **be set** according to the material being machined as shown in Table 11. 1.

切削速度および送りは，表11.1のように，切削する材料によって**設定する**こと.

These operators can also be used to calibrate and zero **set** the probe systems.

これらのオペレータは，プローブシステムの補正およびゼロに**セットする**ことにも使用できる．

Insert the tap in a drill chuck in the tail stock and **set** the lathe on low speed.

心押し台のドリル・チャックにタップを挿入し，旋盤を低速に**セットする．**

Set speeds and feeds on a horizontal milling machine.

横フライス盤の速度と送りを**セットする．**

〈用　語　例〉

king pin inclination set　キングピンの傾き
permanent set　永久変形，永久歪み
permanent set in fatigue (settling)　ヘタリ
set hammer　鋸仕上げハンマ
setting angle　取付け角
setting drawing　据付け図
setting thread plug gage　調整ネジ・プラグゲージ
set pressure of safety valve　安全弁設定圧
set point　設定値
set ring　セットリング，締付リング
set screw　止めネジ，押しネジ，ノブ止めビス，ノブ止め小ネジ
set spring　押えバネ
tool setting (tooling, tool layout)　工具セッティング

set up: to arrange　**取付ける，アレンジする，段取りする**

Set up a machine.

機械を**据付ける．**

We **set** the die stock **up** in a shaper and cut a groove.

型材を形削り盤に**取付け**て溝を切った．

The part to be threaded **is set up** between centers in a chuck, or in a collet.

ネジ切り部品を，チャックまたはコレットにセンタ支持で**取付ける．**

Boring and other operations are frequently done on workpiece **set up** on a face plate.

中ぐりなどの作業は，面板に工作物を**取付けて**することが多い．

Set up a dial indicator <u>on</u> the end of the center to check for runout.

振れをチェックするため，センタの端部<u>に</u>ダイヤル・インジケータを**当てる**（セットする）.

The horizontal mill was easier to **set up** than vertical one and was more accurate.

横フライス盤は，立フライス盤よりも**段取り**が容易で，精度も良かった.

How can a steady rest **be set up** when there is no center hole in the shaft?

軸にセンタ穴がない場合，振れ止めはどう**セット**できるか.

This **set up** let us do punching and forming of a 1/4 inch plate in one operation.

この**段取り**で，1/4インチ板の打抜きと成形が1回でできる.

With the H. machine there is virtually no limit to the variety or types of operations that can be performed **in one set-up.**

H機を使えば，**段取り1回で**できる作業の種類，方式は無限である.

situate : to place or put in a certain position — **置く，取付ける**

A hydraulic servo device **situated** <u>in</u> the hub adjusts the propeller blades; in addition the oil from the servo lubricates the gear and bearings.

ハブの<u>中に</u>**取付けた**油圧サーボ装置で，プロペラの羽根を調節する．さらに，このサーボからの油は，歯車および軸受を潤滑する.

取外す

detach : to release or remove from something else or from a group — **取外す**

The test was stopped periodically every

試験は20,000回毎に定期的に

20,000 times, each test piece **being detached.**

停止し，各テストピースを**取外**した．

Once the cylinder-head retaining nuts and any other components likely to interfere with the removal of the head **have been detached** the head may be withdrawn from the studs.

シリンダ・ヘッドの取外しを邪魔しそうなシリンダ・ヘッドの固定ナット，その他の部品を一度**取外して**しまえば，ヘッドを植込みボルトから抜出すことができる．

To change the grease the fixation bolts are to be unscrewed and the bearing cover is to **be detached** from the gear housing (in the case of gear motor, together with the motor).

グリースを交換するには，固定ボルトをねじ戻して，軸受カバーをギアハウジングから**取外す**ことになる（ギアモータの場合には，モータと一緒に）．

〈用　語　例〉

detachable chain　掛け継ぎ鎖

detachable head (of cylinder)　取外し式シリンダ・ヘッド

disconnect: to break the connection of　**取外す，連結を解く**

Disconnect the plug from the receptacle.

ソケットからプラグを**取外す**．

It will, of cource, be necessary to **disconnect** the universal joint, and the torque tube attachment, when fitted from the rear of the gearbox.

自在継手やトルク・チューブ取付け金具が，ギアボックスの後側から取付けられているときには，もちろん**取外す**（分離する）ことが必要である．

When the gearbox is attached to the clutch housing, it may be necessary either to **disconnect** the gearbox from the bell housing or to remove the housing complete with the gearbox, according to the particular car.

ギアボックスがクラッチ・ハウジングに取付けられている場合は，その車種によって，ギアボックスをベル・ハウジングから**取外す**か，クラッチ・ハウジング全体をギアボックスと一緒に取外すことが必要である.

dismount: to get off or down from something on which one is riding	**取外す**

Never **dismount** an undamaged bearing unless it is absolutely necessary!

絶対必要でない限り，決して無傷の軸受は**取外さない**こと.

If a bearing is to **be dismounted**, it is advisable to mark it to show its relative mounted position- i, e., which section of the bearing was up, which side was "front" etc.

もし軸受を**取外す**ことになった場合には，その相対的取付け位置，すなわち，軸受のどの部分が上か，どちらの側が"前"だったかわかる（示す）ように印をつけるとよい.

Larger bearings can easily **be dismounted** from their sleeves by using an SKF hydraulic nut.

SKFの油圧ナットを使うことによって，大きい軸受ほど楽にスリーブから**取外し**できる.

The advantage of using a sleeve are that the shaft seating does not need such accurate machining and that **mounting** and **dismounting** are considerably facilitated.

スリーブを使う利点は，軸の軸受座があまり高精度な切削を必要としないことと，**取付け**および**取外し**がかなり容易になることである.

remove: to take off or away from the place occupied	**取外す**

Remove nuts, washers, bushings and bolts attaching wing spars to fuselage.

翼桁を胴体に取付けているナット，ワッシャ，ブッシュ，ボルトを**取外す**.

Identify parts for reinstallation on same side if both rotor grips **are removed.**

両方のロータ・グリップ（握り）を**取外す**ような場合には，

同じ側にふたたび取付けられるよう，部品に標識をつける．

A special tool for withdrawing a self-aligning roller bearing inner ring —— the outer ring and rollers having **been removed** first —— is shown in Fig. 19.

自動調心コロ軸受内輪取外し用特殊工具——外輪およびコロは最初に**取外し**ずみ——を図19に示す．

If a cover is installed on nut, it will be necessary to **remove** cover prior to **removing** nut.

カバーがナットの上に取付けられている場合には，ナットを**取外す**前にカバーを**取外す**ことが必要．

The drill **is removed** by tapping a wedge (drift) into the slot in the machine spindle as illustated in Fig. 24.

図24の図解のように，機械の主軸の溝に楔（ドリフト）を軽く叩き入れて，ドリルを**取外す**．

Remove the drill from the socket.

ドリルをソケットから**取外す**．

Operators function is to **remove** finished part from the machine.

作業者の役目は，仕上り部品を機械から**取外す**ことである．

Drain hydraulic fluid by **removing** drain plug from bottom of the reservoir.

リザーバの底から排水栓を**取外して**，油圧液を抜く．

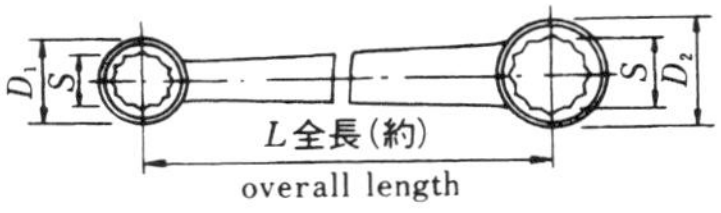

めがねレンチ（offset wrench）

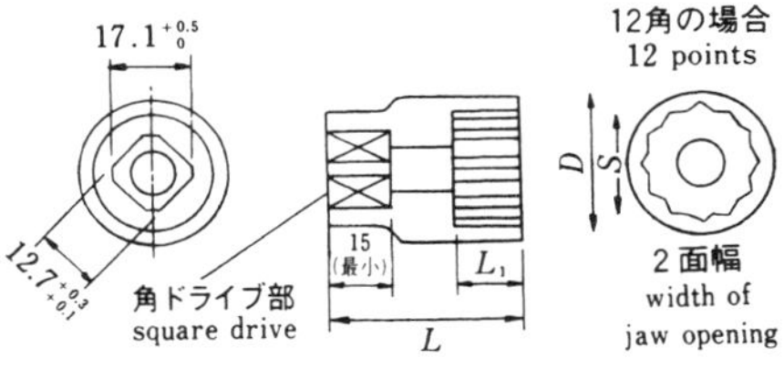

ソケットレンチ用ソケット（socket for socket wrenches）

In difficult cases it may be necessary to heat the inner ring to facilitate its **removal,** but the use of a blow-lamp is out of question unless it does not matter if the bearing is destroyed.

困難な場合には，**取外し**やすくするため内輪を加熱することが必要になる．ただし，軸受が壊れてもかまわない場合以外，トーチ・ランプを使うのは問題外である．

take away: to remove or carry away　　**取外す**

Take away fittings from …….

…から，取付け金具を**取外す**.

take off: to take (clothing etc.) from the body　　**取外す**

When regreasing, the bottom plug **be taken off.**

ふたたび給油するときには，底の栓を**取外す**.

withdraw: to take back or away　　**取外す，引き戻す**

As a rule, there is no difficulty removing **withdrawal** sleeves of the type shown, it being only necessary to tighten the appropriate sleeve nut against the face of the inner ring to **withdraw** the sleeve.

一般に，図示形式の**取外し**用スリーブの取外しで難しいことはまったくない．このスリーブの**取外し**に必要なことは，内輪の側面に，適切なスリーブ・ナットを当てて締めることだけである．

The primary shaft is usually carried in a large double-ball or roller bearing. In most cases on removal of the bearing cover the bearings and constant-mesh pinion **can be withdrawn** without great difficulty.

1次シャフトは，通常大きな複列玉軸受あるいはコロ軸受に組込まれている．多くの場合，軸受カバーを取外すと，軸受および常時嚙合い歯車は，大した困難もなく**取外せる**.

When the tool reaches the end of the workpiece, the lead screw **is** disengaged,

バイトが工作物端に達すると，親ネジの嚙合いがはずれ，バイ

the tool **withdrawn** and the saddle returned to its starting position ready to take another cut.

トは**引っ込み**，サドルはもう一度切削がすぐできるよう，スタート位置に戻る.

The stock pusher advances by an adjustable hydrostatic pressure system and **withdraws** automatically while the last workpiece is being machined.

材料プッシャ（押し）は，調整式静圧圧力システムで前に進む．そして最後の工作物が切削されている間に，自動的に**引っ込む**.

Withdrawing split-pin by levering it against the nut with pair of pliers.

プライヤで割りピンをナットに対してこじって引き出す.

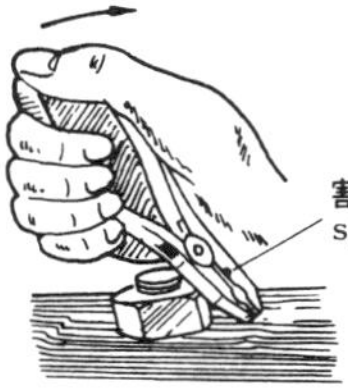

はかる，測る，計る，量る，秤る

determine: to find out or calculate accurately, to detect or calculate precisely

定量する，正確に求める，はっきりさせる

Determine flatness of lapped plate or machine ways with scraped, ground or lapped surface.

ラップ仕上げの板や機械案内面の平坦度を，キサゲ，研削またはラップ仕上げ定盤を使って精密に**求めた**.

Breaking loads of thin-wall iron castings **were determined** with this exper-

肉の薄い鋳鉄品の破壊荷重を，この実験装置で**正確に求めた**.

imental setup.

The width of the wearland can **be determined** by means of a toolmaker's microscope.

摩耗域の幅は，工具顕微鏡で**測る**ことができる．

The tool force **is determined** by measuring the deflection or strain in the elements.

バイト（に加わる）力を，エレメントのたわみまたは歪みを測ることによって**求める**．

A series of tests were run in one system which included 25 pumps and 13 turbines to **determine** how much oil condensed in bearing housings.

軸受ハウジングにどれだけ油が凝縮するかを**はっきりさせる**ために，一連の試験をポンプ25台，タービン13基のシステムで行なった．

The ball separators **were** removed from all test specimens after the life tests and the ball-pocket wear **determined** by a series of measurements of the war scar.

寿命試験後，すべての試験標本からボール分離機を取外して，ボール・ポケットの摩耗を摩耗痕の一連の測定によって**精密に調べた**．

Determined from Co-60 gamma spectrum analysis using NBS SRM 4215-F. Cobalt is present in M50 as a contaminant to a maximum concentration of 2,500ppm.

NBS SRM4215-Fを使い，Co-60ガンマ・スペクトル分析から**正確に求めた**．Coは，M50中に最大濃度2,500ppmまでの汚染要因物として含まれている．

Friction torque, wear, and load-carrying capability **were determined** for the clean bearings and for bearings contaminated with MIL H-5606 paraffinic base, hydraulic oil.

きれいな軸受およびMIL H-5606パラフィン系油圧油で汚れた軸受について，摩擦トルク，摩耗および負荷容量を**測った**．

Friction and wear characteristics **were determined** for loads, temperatures, and

摩擦特性および摩耗特性を，現行の航空機本体用軸受の適用

oscillating conditions that typical of current airframe bearing applications.

個所特有の負荷，温度，揺動条件で**調べた**.

The **determination** of the temperature in the tool in metal cutting…….

金属切削におけるバイトの温度を**正確に求める**ことは…….

〈用　語　例〉
limit of determination　　定量限界

gauge (gage)：to measure exactly　　**測る**

……can **gauge** the amount of oil being fed to the bearing.

……は，軸受に送られている油量を**正しく測る**ことができる.

Convenient visual inspection methods cannot **gauge** the extent of seam or crack in a steel bar.

普通の目視検査法では，棒鋼の亀裂や割れ目の範囲を**厳密に測る**ことはできない.

Do not attempt to rotate the bearing while **gauging** the slackness, but carefully slide the feeler over a roller.

隙間を**測る**間，軸受を回わさずに，フィーラ（隙間計）は1つのコロ上を注意して滑らせるようにする.

The dimension **gauged** extends from the mounting surface on the case to the shoulder of the bearing. The corresponding distance on the transmission assembly is also measured automatically and the two dimensions are compared to determine the shim size required.

測定する寸法は，ケース取付け表面から軸受のショルダー部までである．トランス・ミッション機構のこれに対応する距離もまた自動測定し，この2つの寸法を必要とするシム寸法を決めるために比べる.

Gauging is accomplished through a linear transducer and rotary encoder.

リニア・トランスジューサおよびロータリ・エンコーダで**測る**.

〈用　語　例〉
gage interferometer　　干渉測長器

gauging line of furfural　フルフラール検量線

measure: to find the size or quantity or extent of something by comparing it with a fixed unit or with an object of known size. ; to be a certain size

測る，測定する，計測する，寸法は～である

Rotate hub assembly and accurately **measure** amount of play present.

ハブ・アッシを回わして，そのときの遊びの量を正確に**測る**.

Measure quantitatively the amount of heat given to or taken from the object.

物体に与えられる，あるいは取り去られる熱量を定量的に**測る**.

The instrument generally used for **measuring** temperature is called a thermometer.

温度を**測る**のに，一般的に使われる計器を，温度計という.

The Verimatic Model 8800 automatic hardness tester is an air-operated

8800形自動硬さ試験機は，材料のロックウェル硬さを**測る**，エア作動のコンピュータ化シス

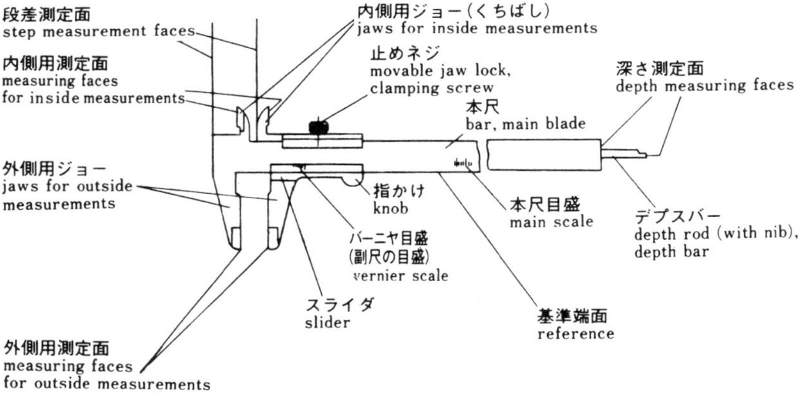

ノギス（vernier calliper）

computerized system that **measures** the Rockwell hardness of materials. | テムである．

These large machines use precision scales to precisely **measure** slide position. | これら大形機械は，スライド位置を精密に**測る**のに精密なスケールを用いている．

Strains are large enough to **be measured** accurately. | 歪みは十分大きいから，正確に**測れる**．

This instrument **measures** basic pitch for spur and helical gears, while they are mounted on the production machine itself. | この計器は，平歯車および傘歯車の基準ピッチを，加工機械に取付けたままで，**測る**ものである．

Wear rates **were measured** for bearings operating under 1,000psi. | 1,000psi で運転した軸受の摩耗速度を**測った**．

The tool forces and chip-thickness **are measured** for each test conditions. | バイト（に加わる）力および切屑厚さを，各試験条件について**測った**．

Remove the bolt and **measure** for the correct length. | ボルトを取外して，長さが正しいかどうか**測ってみる**．

Measure cutting angle to five minutes. | 切削角を5分まで**測る**．

The Zygo bench gage automatically **measures** the diameter or gap to 20 millionths of an inch in approximately one second. | Zygo ベンチ・ゲージは，直径またはギャップを百万分の20インチまで約1秒で自動**測定**できる．

In what units of measure **are** these velocities **measured**? | これらの速度は，どんな測定単位で**測る**か？

B & S measuring microscope **measures** | B&S 測定顕微鏡の測定精度

in ten thousandths.

は，1/10,000である．

Temperature is degree of hotness and coldness **measured** on a definite scale based on some physical phenomenon as the expansion of mercury in a thermometer.

温度とは，温度計の水銀の膨張のようなある物理的現象を基にした特定の尺度で**測った**，寒暖の度合いである．

The volume of the combustion chamber can **be measured** in the way already described.

燃焼室の容積は，既述の方法で**測定**できる．

Measure by the eye.

目**測**する．

……**is measured** with one's eye estimate.

……は，目**測**である．

Measure small length by means of a micrometer.

マイクロメータで，短かい長さを**測る**．

Measure the thickness and the width at several points with a micrometer.

マイクロメータで，厚さと幅を数ヵ所**測る**．

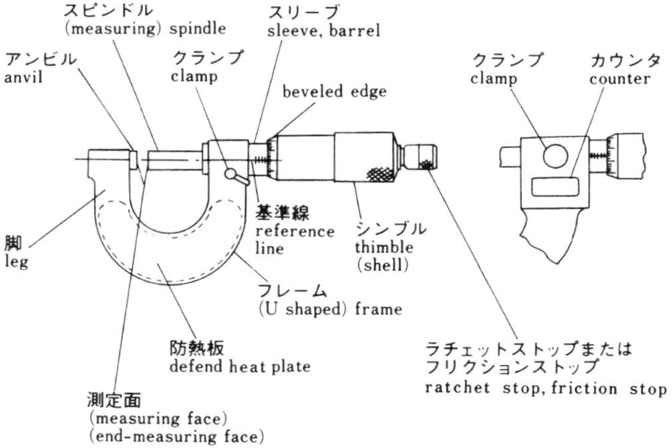

外側マイクロメータ（micrometer calliper for external measurement）

The long diagonal lengths of the indentations **were measured** by a microscope before and after a run.

運転前後に，へこみ（硬さ測定の）の長いほうの対角線の長さを顕微鏡で**測った**．

The distance between the standard markers and the abraded surface **were measured** with a measuring microscope. The volume of material removed was, then, estimated by the average distance.

基準の印とこすり取られた表面との距離を，測定顕微鏡で**測った**．ついで，取り去られた材料の容積を，その平均距離で推量した．

The roughness of the lapped surfaces **measured** by a surface profilometer was 0.2μm for the maximum height of irregularities.

表面プロフィロメータで**測った**ラッピング面の粗さは，凹凸の最大高さが0.2μmであった．

The surface roughness of each section **is measured** with a surface finish measuring instrument.

各部の表面粗さを，表面仕上げ測定器で**測る**．

The impact value **is measured** by the Charpy specimen.

衝撃値は，シャルピ試験片で**測る**．

Conventionally, power **is measured** by the energy which is necessary to raise a given weight to a given hight a given time, 1 footpound per minute of power being the amount of energy required to raise a weight of 1 pound to a height of 1 foot in 1 minute.

普通，仕事率は所定の重さを所定の時間で所定の高さに上げるのに必要なエネルギで**測り**，仕事率1ft・lb/minとは，重さ1lbを1分で高さ1ft上げるのに要するエネルギの量のことである．

It **is measured** by determining the amount of speed increase taking place in a given period of time.

それは，ある所定時間内に起こる速度増加の量を求める（定量する）ことによって**測る**．

The centrifugal separation **is measured** by using Koppers K 36 method, which……

この遠心分離は，Kopper K36法を使って**測り**，…….

The quality of machined surfaces **were measured** using a portable surface indicator.

切削表面の品質は，ポータブル表面インジケータを使って**測った**.

Fuel consumption **was measured** in accordance with the Standard by a volumetric method. That is, the fuel was drawn from a container graduated by 10 ml increments. Total equivalent vehicle kilometers **were** also **measured** for the test using a digital counter on the output shaft.

燃料消費は，規格に従って容量法で**計った**．すなわち，燃料を10 mℓ で目盛った容器から抜き出した．また，この試験では出力軸のディジタルカウンタを使って，等価の車の走行総キロ数を**測った**.

Average grain size **was measured** from scanning electron micrographs.

平均粒度は，走査電子顕微鏡写真で**測った**.

The wear rate of a coupling determines its useful life. An extensive study was performed to determine which factors have the most influence on coupling wear rate. It was first decided to **measure** wear through weight loss.

カップリング（軸継手）の耐用寿命は，その摩耗速度で決まる．どの因子がカップリングの摩耗速度に最も影響するか求めるために広汎な研究を行ない，まず摩耗を重量減により**測る**ことに決めた.

Temperatures **were measured** at five different locations in the system: supply oil temperature, ……, and support bearing.

温度は，このシステムの5ヵ所の違う所で**測った**：給油温度，……，支持軸受.

Measure in the plane containing the direction of primary and feed motion and perpendicular to the direction of feed motion.

主運動および送り運動の方向を含み，かつ送り運動の方向と直角の面で**測る**.

Mesure in the direction of the feed motion.

送り運動の方向で**測る**.

Measure parallel to the feed direction.

送り方向と平行に**測る**.

Measure perpendicular to the plane.	面と直角に**測る**.
Strictly, this dimention should **be measured** both normal to the cutting edge and normal to the resultant cutting direction.	厳密にいえば，この寸法は切れ刃と直角，そして合成切削力方向と直角の両方で**測る**べきである.
Mesurements of a surface **are made** at right angle to the layer.	表面をこの層と直角に**測定する**.
No measurement should **be made** while the cutter is revolving.	カッタの回転中は，決して**測定しない**こと.
Fifty **measurements were made** for each speed of the presses.	プレスの各速度について，50回**測定した**.
The **measurement** of the hardness indentation **is made** with 8 × or 16 × eyepiece at a total magnification of 480 × or 960 × .	8倍または16倍の接眼鏡を使って全倍率480倍または960倍で，硬さ（測定の）へこみを**測る**.
In order to clarify the characteristics and the problems of gas seals, **measurements** of frictional forces and temperatures **were carried out** on the piston-ring type seals with a special experimental apparatus.	ガス・シールの特性および問題点を明らかにするため，ピストン・リング形シールについて，摩擦力および温度を特殊な実験装置で**測った**.
What **are** two common gear **measurements performed** by a mechanist ?	機械技術者が普通行なう2つの歯車**測定**法は何か？
The P 2 has been designed to **take** all **measurements** simultaneously for fast, efficient troubleshooting.	P 2は，故障探求が迅速かつ効率良くできるよう，すべての**測定を**同時に行なえるよう設計されている.
The standard micrometer is used to **take**	この標準マイクロメータは，

measurements <u>to</u> the nearest one-thousand part of an inch.	1/1,000インチ近くまで**測定す**るのに使用する．
Measurement is taken as near to the bearing as possible.	できるだけ軸受に近い所で**測る**．
One should not remove his hand from the handle <u>while</u> **taking measurements.**	**測定**<u>中</u>，ハンドルから手を離さないこと．
Leave it for a while and <u>start</u> **measurements** after the effect of the body temperature is removed and it is convenient to <u>repeat</u> the **measurements.**	しばらく放置し，本体温度の影響を取り除いてから**測定**<u>を始める</u>．そうすれば，**測定**<u>を繰返</u>すのにも都合がよい．
One of the most practical and common **measuring tools** available in the machining and inspection of parts is the steel rule.	鋼製定規は，部品の機械加工および検査で利用できる最も実用的かつ一般的な**測定具**の1つである．
The amplifier has a meter which indicates the position of the **measuring stylus** for measuring full profile.	この増幅器には，全輪郭が測れるよう，**測定針**の位置を示すメータが付いている．
The **measurement** must be between 3.061 and 3.063 inches.	**測定値**は，3.061と3.063インチの間になければならない．
The **measurements** <u>of</u> flank wear-land width are plotted against time.	逃げ面の摩耗域幅<u>の</u>**測定値**を，時間に対してプロットする．
Generally the standard deviation of the mean of thirty holes is ±2 percent and ±10 percent for the individual **measurement.**	全般的には，穴30個の平均の標準偏差は±2％で，個々の**測定値**については±10％である．
Machinability is <u>a</u> **measure** of the ease of cutting of various materials.	被削性は，いろいろな材料の切削難易度の<u>1つの</u>**測度**である．

In experimental work, the depth of the crater is usually a **measure** of the amount of crater wear and can be determined by a surface-measuring instrument.

実験では，普通，クレータ深さはクレータ摩耗量の**測度で**，これは表面測定器ではかることができる．

The objective of this life test program were to (1) obtain an accurate **measure** of gimbal bearing performance, (2) determine the degree of correlation between materials wear tests and full-scale bearing tests and (3) …….

この寿命試験プログラムの目的は，(1)ジンバル軸受性能の正確な**測度**を得ること，(2)材料摩耗試験と軸受現物試験との相関度を確定すること，(3)……である．

The vertical and horizontal components of the deflection on the cantilever under the action of the resultant tool force are taken as a **measure** of the two force components.

バイトの合成力が働いている片持梁におけるたわみの垂直および水平方向の成分を，2つの力成分の**尺度**とする．

Surface roughness **measures** 0.711μm and the visual appearance of the slab was smooth and shiny bright.

測定表面粗さは0.711μm **で**，スラブの外観は滑らかで，ひかり輝いていた．

Workload capacity is 4,400lb and worktable **measures** 19 × 63″.

載せられる工作物は4,400 lbまで，ワークテーブルの**寸法は**19×63インチ**である**．

A miniature clamp **measuring** $2\frac{3}{8}$″ in length and $1\frac{3}{8}$″ in height has been designed for electronic applications.

寸法が長さ$2\frac{3}{8}$インチ，高さ$1\frac{3}{8}$インチのミニチュア・クランプは，電子機器用に設計したものである．

The interior **measured** 17 feet long, 7 feet wide, and 7 feet high.

内部**寸法**は，長さ17ft，幅7 ft，高さ7 ftであった．

The 102-L work tank **measures** 762 ×

102-L 作業用タンクの**寸法は**

406 × 356 mm providing ample room for five parts at once, while its removable front door permits easy loading and unloading of parts.

762×406×356mm で，一度に5個の部品を十分に収納でき，しかも前面のドアは取外しできるので，部品の積み卸しが楽にできる．

〈用　語　例〉

direct measurement　直接測定
measurand　測定量
measured drawing　実測図
measured internal clearance　測定隙間
measuring apparatus　測定器
measuring funnel　計測漏斗
measuring instrument　測定計器
measuring tape　巻尺
measurement　測定，仕上り寸法，測定値
quantity measuring fuse　定量ヒューズ
region of tooth profile measurement　歯形検査範囲
steel tape measure　金巻尺
thread measuring three wire　ネジ測定用三針

meter: to measure by meter　**測る，計量する，量は～である**

With a four-minute lube interval, one of these lube pump can **meter** as low as 473 cm^3 (one pint) per year of oil to a lube point on the compressor.

4分間隔の潤滑で，これら潤滑ポンプの1つは，コンプレッサの1潤滑個所に年473cm^3（1パイント）くらいの少量の供給ができる．

A third approach—microlubrication—provides accurate **metering** of minimum amount of oil required for lublication of the compressor cylinder walls.

第3の方法—ミクロ潤滑—は，コンプレッサのシリンダ壁の潤滑に必要な最少の油量を，正確に**計量する**ことができる．

Two of these lublicators are used to feed oil to the compressor air suction. Each lubricator **meters** only 118 cm^3 over the cource of six month.

これら2つの潤滑器は，コンプレッサのエア吸い込み部に油を送るために使用する．潤滑器はいずれも，半年間の**流量**がわずか118cm^3にすぎない．

The **metering** device : A predetermined amount of lubricant required for the lubrication point **is measured** in the time intervals between applications.

計量装置：潤滑点に必要なあらかじめ決めた潤滑量を，給油間の時間間隔で**計る**.

Analog devices are often used in **telemetry** applications for **measuring** river levels, electrical current, air pollution, and other such continuously variable values.

アナログ素子は，河の水位，電流，空気汚染など連続して変化する値を**測る**ための**遠隔測定**用によく使われる.

〈用　語　例〉

meter rule　メートル尺
metered flow　規制流れ
metering rod, metering pin　測定棒
metering pump, proportioning pump　定量ポンプ
metering screw　計量スクリュ
tele metering　遠隔測定

pick up : to measure　**測る**

Pick up the outside of the solid jaw of the vise body.

万力の固定ジョーの外側を，**測る**.

With an edge finder, **pick up** the edges of the workpiece.

エッジ・ファインダで，工作物の縁を**測る**.

read : (of a measuring instrument) to indicate or register　**読む，読取る，読み出す，示す**

The thermometer **reads** 20°.

温度計は20°を**示す**.

The Angle-Check gauge measures the total included angle or a taper in less than two seconds. It **reads** directly in degrees and minutes, requires no calculation, …….

Angle-Check ゲージで，角度やテーパを２秒以内で測る．度と分で直接**読取れる**ので，まったく計算を必要とせず，…….

The difference in measurements can **be read** by resetting to zero at any point in

この範囲内のいずれの点で，ゼロにリセットしても，測定値

the range.

の差を**読取る**ことができる.

The height of the meniscus **is read** on a millimeter scale.

凹凸の高さをミリメータ目盛で**読む**.

Instrumentation. The measuring cycle is controlled and recorded with a programmable gauging system. This equipment includes a Hewlett-Packard 9825 A computer which is programmed via cassette tapes to take **readings** at specific points in the cycle.

計装. 測定サイクルはプログラマブル・ゲージング・システムで制御および記録する. この装置は，サイクル中の指定個所で**読取れる**ようにカセットテープにプログラムした Hewlett-Packard 9825A コンピュータを備えている.

〈用　語　例〉

read group total　グループ合計点検
reading　読み，読取り
read-only store　読取り専用記憶装置
read/write head　読取り・書き込みヘッド

---

scale: to represent in measurements or extent in proportion to the size of the original　**測る**

---

scale the drawing.

図面の寸法を**測る**.

---

size: to group or sort according to size　**寸法決めする**

A. M. in-process, double finger, automatic gage **sizing** unit is used throughout this operation. The gagehead continuously measures the bearing race during grinding, sending back a signal to initiate rapid wheel head retraction the instant required size is reached.

この作業は，始めから終わりまで，A.M.インプロセス・ダブルフィンガ自動**定寸**装置が使われる. ゲージ・ヘッドは研削中ベアリング・レースを連続測定し，必要とする寸法になった瞬間，砥石台を急速後退させるよう信号を送り返す.

The most important factor when selecting a mist generator head is not the type but the size. A mist generator head must **be sized** to match its application.

ミスト・ジェネレータ・ヘッドを選定するときの最も重要な条件は，形状ではなく寸法である．ミスト・ジェネレータ・ヘッドは，その用途にマッチするよう**寸法を決め**なければならない．

〈用　語　例〉

automatic sizing device　自動定寸装置　　sizing reel　検尺器（ラップリール）

---

survey: to measure and map out the size, shape, position, and elevation etc. (of an area of the earth's surface)

実測する，測量する

……**make a survey** of……

……を**測る．**

〈用　語　例〉

surveyed drawing　実測図

surveying laser product　測量用レーザ製品

---

take: to find out and record → **measure**

～を測る

Take temperature.

温度を**測る．**

Take your measurement.

君の寸法を**測る**（とる）．

A series of carefully **taken** measurements……

慎重に**測った**（**とった**）一連の**測定値**は……．

---

weigh: to measure the weight of, especially by means of scales or a similar instrument

（重量を）計る，秤る，重さは～である

Weigh 100 mg accurately, put in a 200 mℓ measuring flask and make it to 200

100mg を正確に**計り**，200mℓの測定フラスコに入れて，水

mℓ with water.

を加えて200mℓにする.

Weigh out accurately 12.5 mg of amine.

アミン12.5mgを正確に**計りとる**.

Any production worker can master the operation of a S. counting system in seconds. Just **weigh** the reference parts, set the reference indicator, and start **weighing.** Read the count on the R in seconds.

どんな作業者でも，S計数方式の操作を数秒でマスターできる．まず基準部品を**測り**，基準インジケータをセットし，**測定**開始．一瞬のうちに，Rの数を読む.

Weigh the parts on a S. electronic scale that is connected to an R. part counter.

R部品計数器に連結したS電子スケールで，部品重量を**計る**.

S. has systems that can count quantities up to 99.999 and parts that **weigh** as little as 10 mg.

S社には，**重量**10mgという軽量部品を数量99.999まで計数できるシステムがある.

The robot can handle parts **weighing** up to 20kg….

このロボットは，**重量が**20kgまでの部品を取扱うことができ，…….

The machine operates at 20 spm, has a stroke length of (660mm), **weighs** more than 240,000kg and measures 6m wide and 6.70m long.

この機械は20spmで運転され，ストローク長さは660mm，**重量**240トン以上，寸法は幅6m，長さ6.70mである.

〈用　語　例〉

weighing bottle　秤りびん　　weighing capacity　秤量

はかる，図る，もくろむ（計画，設計，立案）

arrange: to form plans	計画をつくる

If control of the two horizontal axis of the table supporting the workpiece **is arranged**, the machine can be programmed to locate and then drill a specified pattern of holes.

工作物の支持テーブルが2水平軸制御に**設計されて**いれば，機械は指定穴パターンを位置決めして穴明けするようプログラムできる．

contrive: to plan cleverly	上手に計画する

Piston is a cast circular piece of metal so **contrived** that it works with an oscillating or up-and-down movement within a hollow cylinder.

ピストンは，中空円筒の中で往復，即ち上下運動で働くように，**うまく設計された**金属の鋳造環状部品である．

design: to plan	計画する，設計する，デザイン

Like all machine tools, the C/C **is designed** and built to the highest standards of quality, with induction-hardened headstock gears, and precision Gamet spindle bearings.

すべてのC工作機械同様，C/Cは最高規格の品質に**設計**・製作されており，高周波焼入れ主軸台歯車・精密G主軸軸受を用いている．

We can custom **design** and manufacture to meet your particular requirements,

特定の要求条件に合わせて，受注**設計**ならびに製作できる．

The company **redesigned** basic castings **to** make them easier to machine.

加工しやすいように，鋳物の原形を**設計し直した**．

The machine **is designed** to remove more metal at faster speeds with less power

この機械は，少ない動力消費で，より高速で，より多く削れ

consumption,

るように**設計されている.**

A noncontact scanning laser beam gage, small enough to fit on a bench top, **has been designed** to eliminate the human error factor associated with conventional gaging methods.

机上に十分取付けできるぐらい小さい，非接触走査レーザ光線ゲージは，普通の寸法測定方式に伴う人的誤差因子をなくするように**設計されている.**

The control **is designed** to permit programming by inexperienced opreraters without having to consider the tool path.

この制御装置は，未経験のオペレータがツール・パスを考えることなく，プログラミングできるように**設計してある.**

The rugged jamieson JD 6 Cam Unit **is designed** for easy, problem-free operation with a high degree of reliability.

この頑丈なJD6カムユニットは，信頼性が高く，容易かつ問題なく運転できるよう，**設計されている**…….

The Mega 75 shaft drilling system **designed** for drilling deep holes in cylindrical or long rectangular parts or for drilling shallow holes, can drill up to a 45″ hole in one pass. **Designed to** hold size, straightness, roundness and accuracy, hole drilling capacity ranges from 0.0781″ to 1.000″ in diameter depending on application.

円筒または長い矩形部品に深い穴をあけ，または浅い穴をあけるために**設計された**このM75軸穴明けシステムは，1パスで45インチの穴をあけることができる．寸法，真直度，真円度および精度を維持するように**設計されて**いて，穴明け能力範囲は，用途によるが直径で0.0781～1.000インチである.

The collet-type straight shank tool holders **are designed** for any machine with a hollow spindle.

コレット形直線シャンク・ツール・ホルダは，中空主軸のどんな機械にも使えるように**設計してある.**

Ball and roller bearings must **be designed** for actual life requirements

玉軸受およびコロ軸受は，実際の寿命必要条件（通常数10^6

(usually millions or billions of cycles) and they can not **be designed** for infinite life.

または数10^9サイクル）に合うよう**設計**しなければならない．そして寿命が無限になるようには**設計**できない．

Both series **are designed** for use on conventional machine tools and machining centers and **are designed** to provide greater economy in production applications.

両シリーズとも，普通の工作機械およびマシニングセンタに使うように**設計され**，経済的な生産が行なえるよう**設計されている**．

The SAFI (Servo Actuators For Industry) line **has been designed** for use in robotic and factory automation applications to meet a variety of industrial automation requirements from complex robots to simple material handling and parts positioning devices. The actuator units **are designed** to provide precise rotary motion and act as a self-contained hinge point for articulated robotic joints.

このSAFI（産業用アクチュエータ）系列は，複雑なロボットから簡単なマテハンおよび部品位置決め装置に至るまで，各種自動化の要求条件に合い，ロボット化および工場自動化用に使えるよう**設計されている**．このアクチュエータは，精密な回転運動を与え，かつロボットの関節の内蔵ヒンジ部として働くように**設計されている**．

This feature allows users of CAD/CAM systems to **design** parts on the screen and immediately produce a tool tape for the NC machines.

このような特徴があるので，CAD/CAM システムのユーザーは，スクリーン上に部品を**設計し**，即座にNC機用ツールのテープをつくることができる．

To **design** a machine around a cutting tool, however, requires accurate knowledge of the potential performance of the tool.

ある切削工具を中心にして機械を**設計**する場合には，その工具の潜在能力についての正確な知識が必要である．

Fixturing should **be** specifically **designed** around the peculiarities of the workpiece to be carried.

取付け方式は，搬送する工作物の特異性を重点に，特別な**設計をする**．

The Series BP blanking and progressive die press line customized for greater productivity in the 100 to 800 tonnage range, **is designed for** autmatic production utilizing various types of feeds or transfer mechanisms. Feeding and quick die-changing systems **have been designed** so that short run production may be automated.

100～800トンの範囲で，より生産性を高めるために受注製作したBD打抜き，連続ダイプレス・ラインは，各種の送り方式や移送機構を利用して，自動生産用に**設計されている**．送りおよびクイック・ダイ・チェンジシステムは，少量生産の自動化ができるように**設計されている**．

The tools **are designed** with a split point which self-centers for hole location accuracy. The pilot diameter, then **is designed** to ease penetration and stabilize the tool in the part, to help allow a cleaner, freer cutting action by the drill diameter.

このツールは，穴位置の精度が出るように，みずからセンタを出す割り形先端に**設計されている**．さらに，パイロット径は，ドリル径でキレイで楽な切削ができるよう，入りやすくかつ部品の中で安定するように**設計されている**．

Computer **designed** cast-iron construction maintains rigidity, vibration damping and thermal stability.

コンピュータ**設計**の鋳物構造であるから，剛性，振動の減衰，熱安定性が保てる．

The sturdy **design**, employed throughout the machine, enabies accomodation of workpiece up to 2.000lb.

機械のすみずみまで頑丈な**設計**が採用されているので，2,000 lbまでの工作物が扱える．

Single belt unit can transmit up to 50 hp, with capacities up to 100hp being offerred **in** double belt **design.**

1本のベルト・ユニットは50馬力まで伝達でき，2本ベルトの**設計**のものには容量100馬力までのものがある．

Lacking a suitable guide to the solution of grease flow problems, the tendency is to over- or under- **design**.

グリースの流れの問題を解くのに適切な手引きがないと，過剰**設計**あるいは過少**設計**になりやすい．

engineer: to construct or control as an engineer to contrive or bring about (to plan clevery, to manage)	**設計する，処理する**

How to **engineer** a centralized lubrication system.

集中潤滑方式の**設計製作**の仕方.

Properly **engineered**, operated and maintained, centrifuges will remove all except trace amounts (20 ppm) of free water from turbine lube oils.

操作および保全が適切に**設計製作された**，遠心分離機は，タービン潤滑油から遊離水の痕跡量（20ppm）以外すべて除去する.

The massive bidirectional turret **is** precision-**engineered**, mounted on a slant saddle, rigidly supported and unaffected by any heavy tool imbalance.

どっしりした２方向タレットは，傾斜サドルに取付けられ，頑丈に支持され，ツールがどんなにひどくバランスを失ってもその影響を受けないように，精密に**設計製作され**ている.

Whether you need a component or a complete plant, Paxson will **engineer** it to meet your needs.

Paxson 社は部品であれ工場全体であれ，必要に合うよう**設計製作**する.

A line of sliding headstock automatics **is engineered** to hold extremely close tolerances in the production of precision components from bar stock.

主軸台移動形自動機のシリーズは，棒材から精密部品を生産する場合，きわめて狭い公差を保つよう**設計製作されている**.

plan: to arrange a method etc., for to make a plan or design of	**計画する**

Plan a fully-automatic production line.

全自動生産ラインを**計画する**.

If this is done, good record of the bearing will be obtained and any future replacement can **be planned** well in advance.

こうすれば，軸受について良い記録が得られ，将来のどんな交換もあらかじめうまく**立案**できる．

Where there is a possibility of overhead drip, **plan** on protecting the pump and its drive with some type of drip pan.

頭上から滴れそうな場合は，何らかの形式のしずく受け皿でポンプとその駆動装置を保護するように**策を立てる**．

Every detail of system enginnering **is planned** for positive protection of machine mechanism for operator safety and convenience.

システム・エンジニアリングの細目は，機械機構の確実な防護とオペレータが安全で使いやすいように，**設計されている**．

An R & D **plan** in tribology, when realized and implemented, offers a potential saving of 11 percent of total u. s. energy consumption.

トライボロジーの研究開発**計画**は，具体化されて実施されたときには，アメリカの全エネルギ消費量の約11%を節約できる可能性を秘めている．

program (programme): a plan of intended proceedings

プログラムを組む，プログラム，実行可能プログラム，ルーチン

This can **be** accomplished manually or **programmed**.

これは手動でも，**プログラム**でもできる．

The minicomputer-based CNC control permits shopfloor **programming** and editing.

このミニコンピュータ・ベースのCNC制御装置で，工場現場で**プログラミング**と編集ができる．

With machines incorporating feed back control, **program** can be provided in the

フィードバック制御の機械は，**プログラムを**，ディスクおよび

form of punched tape or punched cards which are inexpensive to produce compared with disc and drum cams.

ドラム・カムに比べ費用のかからない紙テープまたは穿孔カード方式で，作れる．

schedule: a programme or timetable of planned events or of work, to include in a schedule, to appoint for a certain time

計画する，日程計画，（指定）工程

The computer **schedules** an NC gantry robot which loads machine tool changers as needed for job orders and replaces worn or broken tools.

このコンピュータは，仕事に必要な順序に従って，機械の工具交換装置に搭載したり，摩耗あるいは破損したツールを交換するNCガントリー・ロボットの**手順を決める**．

It is equally important to be able to calculate the effects of delays and to **reschedule** as the project progresses.

計画の進行とともに，遅れの影響を計算して**日程を組み直し**できることも，同じように重要である．

High quality jobs **schedule** on the presses average 220,000 pieces, but may range from a low of 25,000 to a high of between two and three million.

これらのプレスでの質の高い作業の日程**計画**は平均220,000個で，少なくて25,000個，多くて2～3,000,000個の間である．

Sixty-nine automatic pressworking lines **are scheduled** to be built in 1978.

69の自動プレス作業ラインは，1978年につくるよう**計画されている**．

Separate magazines may **be programmed** in or out of the working position, depending upon the workpiece that **is scheduled** for the individual machining center.

個々のマガジンは，各マシニングセンタに工程が組まれている工作物に即して，作業位置への出・入を**プログラムできる**．

挟む，くわえる

nip: to pinch or squeeze sharply, to bite quickly with the front teeth	挟む，くわえる

The gudgeon pin is held in place by means of two circlips fitting in the grooves just inside the pin holes in the piston. The ends of these can **be nipped** together by means of fine-pointed pliers and the clips removed.

ピストンピンは，ピストンのピン穴のすぐ内側の溝にはまる2つのサークリップによって，所定位置に保持される．サークリップの端を，細く尖ったプライヤによって，**挟みつけ**てクリップを取外すことができる．

pinch: to squeeze tightly or painfully between two surfaces, especially between finger and thumb : as much as can be held between the tips of the thumb and forefinger	挟む，つねる

In using the inlet valve for retiming, the feeler **is pinched** <u>between</u> tappet and valve at the correct moment of opening and released when it closes.

タイミング再調整に吸入弁を使う場合には，すきみ計はバルブが正常な開きのときにタペットと弁の<u>間に</u>**挟まれ**，それが閉じるときに解放される．

Insert a feeler guage 0.016in. thick between tappet and valve, and turn the engine until it **is pinched by** thc rising tappet.

0.016インチのすきみ計をタペットと弁との間に挿し込み，それが上がってくるタペットに**挟まれる**までエンジンを回わす．

Fingers are interchangeable with internal gripping, external gripping and soft

フィンガは，内つかみ，外つかみ，およびソフト・ブランク

blank options. Two-or three-finger configurations operate on plant air, where pressure acts to open or close fingers. **Pinch force** is proportional to inlet pressure and varies between 2 lbs. and 55 lbs.

用（オプション）と互換性がある．2指または3指構成のものは，工場内のエアで作動する．この場合，圧力で指を開いたり閉じたりする．**挟持力**は入力圧に比例し，2～55ポンドの間である．

ひく

sawing : cutting off with saw　　**鋸引き**

Sawing is used mainly for cutting off and for cutting plate material of not too great a thickness.

鋸引きは，主に切断およびあまり厚くない板材を切るのに使われる．

Saw it in half lengthwise.

長手方向，その半分に**鋸でひく**．

分解する

break up : to separate　　**分解する，解体する**

Break up the things which are fastened.

固定してあるものを**分解する**．

disassemble : to separate　　**分解する**

Without **disassembling** the machinery, a mechanic or technician can use a direct reading F-graph on site to quick screen lubricant samples.

機械工や専門技術者は，機械を**分解**することなく，潤滑剤試料の迅速な篩分けに現場で直読式Fグラフを使うことができる．

Since the AC servo is preventive-maintenance-free, it is best suited for this application with need to regularly **disas-**

ACサーボはメンテナンス・フリーであるから，サーボの周期的保全のために機械を定期的

semble the machine for periodic servo maintenance.	に**分解する**必要のあるこの用途に最適である.

dismantle : to take away fittings or furnishings from, to take to pieces	**分解する**

Secure the connecting rod to the pin and give the shaft a turn or two and **dismantle** it again.	コンロッドをクランクピンにしっかり取付け，クランクシャフトを1～2回回わして，ふたたび**分解する**.
After 300, 1,000, 3,000, 10,000, and 25,000 hours, the bearings **are dismantled** for observation of the components and the lubricant : once the observation is accomplished, the bearings are assembled again, without replacing the lubricant.	300，1,000，3,000，10,000，25,000時間後に，部品および潤滑剤を観るために軸受を**分解する**．観察し終わったら潤滑剤を変えないで，軸受をふたたび組立てる.
On **dismantling** of temporary constructions, the contacting surfaces are easily damaged.	仮組みしたものを**分解する**ときは，接触面を傷付けやすい.
The **dismantling** of the drive depends on the type of drive.	駆動装置の**分解**の仕方は，その駆動形式によって違う.

take apart : to separate	**分解する**

To **take apart** the pump body see section 5.1.	ポンプ本体の**分解**には，5.1項を見よ.
Bench vises **should** occasionally **be taken apart** so that the screw, nut, and thrust collars may be cleaned and lubricated. The screw and nut should be cleaned in solvent.	卓上バイスは，ネジ，ナット，スラスト・カラーをきれいにし，そして潤滑できるよう，ときどき**分解する**こと．ネジとナットは溶剤の中で洗浄すること.

磨く

polish : to make or become smooth and glassy by rubbing	**磨く，研磨，研削**

Polish well to make the surface flat.

表面を平らにするようによく**磨く**.

Minor nicks, scratches, and gouges may be **polishd** out. Blend edges of repaired area into surrounding surface with a smooth contour.

小さい欠け傷，掻き傷およびかじりは**磨いて**とる．直した部分の縁をその周りの表面と滑らかな輪郭になるように丸める.

Polish steel parts only to a depth sufficient to remove traces of corrosion.

鋼の部品は，腐食の痕がとれるだけの深さまで**磨く**.

For **polishing** with abrasive cloth, set the lathe for a high speed and move the cloth back and forth across the work. Hold one of the cloth strip in each hand.

研磨布紙で**磨く**には，旋盤を高速にセットして，研磨布紙を工作物の端から端まで前後に動かす．研磨布片の端をおのおの手で持つ.

Polish to remove corrosion pitting on the outer 2/3 of the vane.

羽根の外側2/3の腐食痕を除去するために**磨く**.

〈用　語　例〉

electro-polishing　電解研磨
polisher　磨き皿
polishing　つや出し，バフ加工
polishing machine　バフ研磨機
polishing powder　微粉

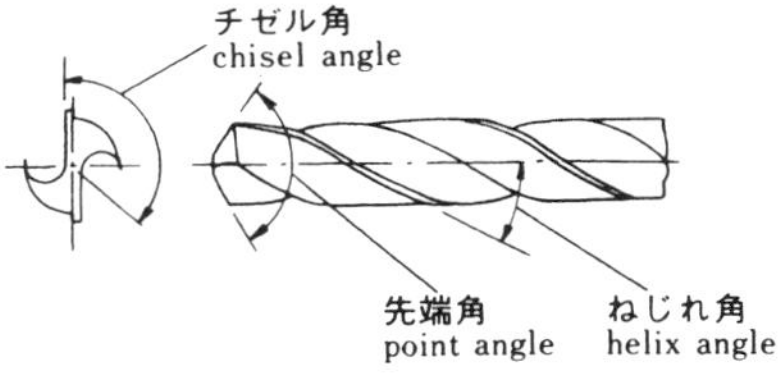

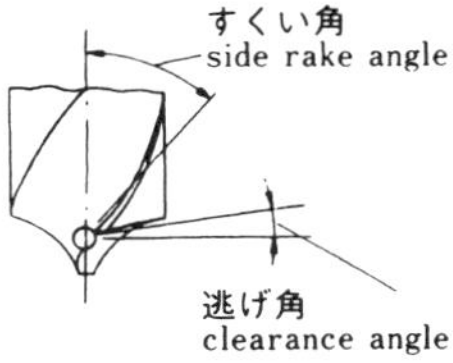

sanding : to smooth or polish sand or sand paper	磨く，研磨

Let dry until not tacky and lightly **sand** with No. 400 paper and wipe with a cloth.

ねばつかなくなるまで乾かして，No. 400のペーパーで軽く**磨き**，布で拭う．

Sanding shall be accomplished by hand only.

（ペーパー）**磨き**は手ですること．

〈用　語　例〉

belt sanding　ベルト研磨

water proof sand paper (glass paper)　耐水性研磨紙

見付け出す（検知，検出）

ascertain : to find out by making enquiries	調べる（検知）

Before attempting to turn a screw it should **be ascertained** that the screwdriver blade is well down in the screw slot. If it rides on the edge through being too blunt, it will injure the slot.

ネジを回わそうとする前に，ネジ回しの刃がネジのすり割に十分底まで入っていることを，**調べる**こと．もし刃先が丸くなっていて，端に乗り上がると，すり割りを傷付けてしまう．

Long-term testing is required to **ascertain** whether deleterious or beneficial effects of this kind exist.

この種のものの効果が良いか悪いかを**見極める**には，長期にわたって試験することが必要である．

Dipstick. ――A narrow metal rod which, being dipped into a petrol-or oil-tank, is used to **ascertain** the depth of fluid therein.

検油棒――ガソリンタンクや油タンクの中に漬けて，その中にある液の深さを**確かめる**のに使う，細い金属の棒．

Ascertain by inspection.	検査して**見付け出す．**

detect: to discover the existence or presence of	**見い出す（発見，検出，検知）**

Detect a weak spring by inserting a screw driver in the coil.	コイルの中にネジ回しを挿し込んで，弱いバネを**検出する．**
A force can **be detected** by the gauges.	力はゲージで**検知**できる．
The variations **are detected** with a current transformer which relays the effect to a multimode process controller.	この変化は，その結果をマルチモード・プロセス制御器に中継する変流器で**検知する．**
Where evaporation loss has been eliminated, as in sealed foil envelopes, lubricated bearings are stored for many months without **detectable** loss of this additive.	潤滑してある軸受は，蒸発がないようにした（密封した箔の包の中のような）場合には，長い月日保管しても，この添加剤の喪失は**検知でき**ない程度である．

〈用　語　例〉

error detecting code　誤り検出コード

undetective error rate　見逃し誤り率

gas detection tube　ガス検知管

determine: to find out or calculate accurately : to detect or calculate precisely	**調べる**

The test fixture was monitored hourly with a stethoscope to **determine** when rolling contact fatigue failure had occurred.	転り接触疲労損傷の発生したときを**見い出す**ため，試験取付け具を聴診器で1時間ごとにモニターした．

Efficiencies of recovering oil from the resin, **were determined** for three different oils commonly used in cold rolling foil or sheet.

普通，箔や薄板の冷間圧延に使われている3つの異なった油について，レジンからの油回収効率を**求めた．**

〈用　語　例〉

limit of determination　　定量限界

trace: to follow or discover by observing marks os tracks or pieces of evidence	**追跡調査する**

Water is also an electrical conductor. This fact should be borne in mind when **tracing** electrical faults due to short circuits in the electrical system.

水もまた電気の伝導体である．電気系統の短絡による電気的故障を**追跡調査する**ときには，このことを念頭におくこと．

The variation in each bead seat **is traced** <u>by</u> a mechanical probe which physically contacts the rim.

各ビードシートの変動量は，リムに物理的に接触する機械的プローブ<u>によって</u>**追跡する．**

Sometimes the rattling can **be traced to** the corners of the bonnet, …….

ときに，ガタガタする源を**追跡してみると**，ボンネットの隅ということがあるが，…….

Fully 90 percent of all fatigue failures occurring in service or during laboratory and road tests **are traceable** <u>to</u> design and production defects.

実用中，あるいは研究所および路上試験で発生する全疲労損傷の90%を**たどってみると**，全部が設計および製作の欠陥<u>で</u>ある．

〈用　語　例〉

tracing program　　追跡プログラム
tracing routine　　追跡ルーチン
traced drawing　　トレース図
tracer　　トレーサ，写図工
tracer controlled lathe　　ならい旋盤
tracing cloth　　トレース布
tracing paper　　トレース紙

trouble shoot: to trace and correct faults in machinery etc.	**調べて直す（故障探求）**

……be checked first when **trouble-shooting** this system.

このシステムの**故障を探求する**ときには，まず……をチェックする．

Cycle counters are aids in **trouble-shooting**.

サイクルカウンタは，**故障探求**に役立つ．

持つ，保つ，保持する，維持する

hold: to take and keep in one's arms, hand (s), teeth, etc. : to grasp or keep so as to control : to have or keep as in a grasp : to support, to bear the weight of	**持つ，つかむ，支える，保つ**

Hold the chisel firmly enough to guide it, but the same time light enough to ease the shock of the blows.

たがねは，それないようにしっかり持ち，同時にハンマーの打撃ショックを和らげられる程度に軽く**持つ**．

Withdrawal of the paper should be done rapidly. The paper should **be held** at the lefthand corner between the thumb and the first finger of the left hand and lightly pulled at the same time as the paper release lever is operated. The paper can thus be removed quickly from the machine.

紙を素早く引き出すこと．紙の左隅のところを左手の親指と人指し指の間に**挟み**，紙解放レバーを操作すると同時に，軽く引っ張ること．これで，紙を機械から素早く取外すことができる．

Hold the test-tube with the open end firmly between the first and second fingers.

試験管の開口端を，人指し指と中指の間でしっかりと**挟む**．

If a weighted stick or pendulum **be held** in one hand the stick may be made to swing slowly from side to side with the other hand.

もし，重錘を付けた棒，すなわち振子を片手に**持て**ば，他の手で棒をゆっくりと左右に振ることができる．

Surface heat dissipates so quickly that a tile can **be held** with the bare hand only seconds out of the oven and while the tile's interior is still red hot.

表面の熱が非常に速く放散するから，炉から取り出してわずか数秒，つまり，タイルの内部がまだ赤熱していても，タイルを素手で**持つ**ことができる．

Detachable tappet blocks, if fitted, **should** be removed, or each tappet **held up** by hand clear of the cams as the camshaft is withdrawn.

取外しできるタペット・ブロック（もしついていたら）を取外すこと．あるいはカム軸を引き出すときにカムに触れないように，タペットを個々に手で**持上げる**こと．

Solder …… while **holding** it with a tweezers.

それをピンセットで**挟んだ**ままで，……をろう付けする．

Hold the belt ends wide enough apart to allow the portion of the belt passing round the crank-pin to make contact with an area equal to no more than a quarter of the circumference.

クランクピンの周りを通るベルトの部分が，円周の1/4を超えない領域と接触するように，ベルトの端を十分広く離して**持つ**．

Take long, flowing strokes with the brush, **holding** it so that it is inclined at about 30 degrees to the surface being treated and working in the same direction throughout.

ブラシを，処理する表面に約30°傾けるように**持って**，始めから終わりまで同じ方向で作業し，ブラシを長く流れるようになでる．

The brushes can be removed for examination by **holding up** the pressure spring.

圧力バネを**持上げ**れば，ブラシを調べるために取外すことができる．

Others range from relatively inexpensive **hand-held** models to a ＄180,000 machine.

ほかに，比較的安価な**手持**形から，＄180,000の機械まである．

A clean spanner will not slip like an oily one, which also **gives a** slippery **hold** for the hand.

きれいなスパナは油の付いたもののように滑らない．また油の付いたものは，**手で持っても**滑りやすい．

Vises are used to **hold** work for filing, hacksawing, chiseling, and bending light metal. They are also used for **holding** work when assembling and disassembling parts.

バイスは，軽金属にヤスリがけ，鋸切り，はつり，および曲げたりするとき，ワークを**保持する**ため使われる．これはまた，部品を組立および分解するときに，ワークを**保持する**のにも使われる．

Workpieces can **be hold** in position by means of collets, a 3-jaw chuck or special hydraulic clamping mandrels.

工作物は，コレット，三ツ爪チャックまたは特殊な油圧クランピング・マンドレルなどによって，所定の位置に**保持すること**ができる．

Work holding is accomplished either by using a machine vise bolted to the worktable or by direct bolting of the workpiece onto the worktable using the T slots provided.

加工物の保持は，ワークテーブルにボルト留めした機械バイスを使うか，あるいは既設のT溝を利用して，工作物をワークテーブルに直接ボルト留めするかのどちらかの方法で**できる**．

Work holding is often **achieved** by use of a magnetic vise (controlled by a lever) that can be placed in the "on" or "off" positions.

「オン」，「オフ」位置に位置決めできる磁気バイス（レバーで制御）を使うことによって，**加工物を保持できる**場合が少なくない．

The most widely used type is the center

最も広く使われている機種は

lathe, also known as the engine lathe, in which the work **is held** between centers or in chuck.

センタ・レース（エンジン・レース〈普通旋盤〉ともいう）で，これでは工作物を両センタまたはチャックに**保持する**.

The shaft **is** securely clamped to the bench or **held** in vice.

軸は，しっかりとベンチにクランプするか，またはバイスに**固定する**.

The shaft **is held** mechanically (fixed) in a holder which transmits the rotational motion.

この軸は，回転運動を伝えるホルダに，機械的に**保持**(固定)されている.

The piston ring **is held on** the worktable by a magnetic chuck.

磁気チャックで，ピストンリングをワークテーブルに**保持する**.

Stout and softish brown paper is most suitable, and this can very easily by cut to the proper shape by **holding** it **against** the bottom of the casting and striking the edges with a mallet or a hide hammer.

丈夫で軟かめの褐色紙が最適である．そして，これを鋳物の底に**当てて保持し**，その縁をマレットか皮革ハンマーで叩くことによって，適切な形に，きわめて容易に切ることができる.

If a light-sensitive pen **is held against** the screen, it redisters a signal as the beam passed by.

感光ペンをスクリーンに**当てる**と，ペンはビームがそばを通過すると信号を表示する.

Hold with the bevelled side **against** the floor.

開先した側を床に**当てて保持する**.

〈用　語　例〉

holding down bolt　据付けボルト
holding drum　支持ドラム
holding function　把持機能
holding power　保持力
tool holder　ツールホルダ
work holder　工作物保持具

〔Index〕

〔A〕

〔B〕

〔C〕

〔D〕

〔**L**〕

〔**M**〕

〔**N**〕

〔**O**〕

〔**P**〕

〔**R**〕

〔**S**〕

著　者　略　歴
野澤　義延（のざわ・よしのぶ）［故人］
1939年　東京高等工芸学校精密機械科卒業
1954～1974年　光洋精工株式会社技術サービス部勤務
この間，海外文献の翻訳を手掛ける．

機械を説明する英語

2011年6月20日　第1版第1刷発行
2021年9月10日　第1版第6刷発行

著　　者　野澤義延
発 行 者　森北博巳
発 行 所　**森北出版株式会社**
東京都千代田区富士見1-4-11（〒102-0071）
電話 03-3265-8341／FAX 03-3264-8709
https://www.morikita.co.jp/
日本書籍出版協会・自然科学書協会　会員
JCOPY ＜(一社) 出版者著作権管理機構 委託出版物＞

落丁・乱丁本はお取替えいたします　印刷・製本/創栄図書印刷

Printed in Japan／ISBN978-4-627-94591-3